AN ESSAY

ON THE

STEAM-BOILER.

READ BEFORE THE

FRANKLIN INSTITUTE, PHILADELPHIA, JANUARY 16, 1867.

BY JOSEPH HARRISON, JR.,
MECHANICAL ENGINEER.

TO WHICH IS ADDED

A REPORT OF THE COMMITTEE ON SCIENCE AND ARTS, CONSTITUTED BY THE FRANKLIN INSTITUTE, ON THE

HARRISON STEAM-BOILER.

REPORT OF BOILER EXPLOSIONS IN ENGLAND FOR 1866.

BY E. B. MARTEN,
ENGINEER OF MIDLAND BOILER ASSOCIATION,

ETC., ETC.

MAY, 1871.

HARRISON BOILER IN COURSE OF ERECTION.

VIEW OF BOILER WORKS.

HARRISON BOILER.

SAFE.

AN ESSAY

ON THE

STEAM-BOILER.

READ BEFORE THE

FRANKLIN INSTITUTE, PHILADELPHIA, JANUARY 16, 1867.

BY JOSEPH HARRISON, JR.,

MECHANICAL ENGINEER.

TO WHICH IS ADDED

A REPORT OF THE COMMITTEE ON SCIENCE AND ARTS, CONSTITUTED BY THE FRANKLIN INSTITUTE, ON THE

HARRISON STEAM-BOILER.

REPORT OF BOILER EXPLOSIONS IN ENGLAND FOR 1866.

BY E. B. MARTEN,

ENGINEER OF MIDLAND BOILER ASSOCIATION,

ETC., ETC.

MAY, 1871.

DANGEROUS.

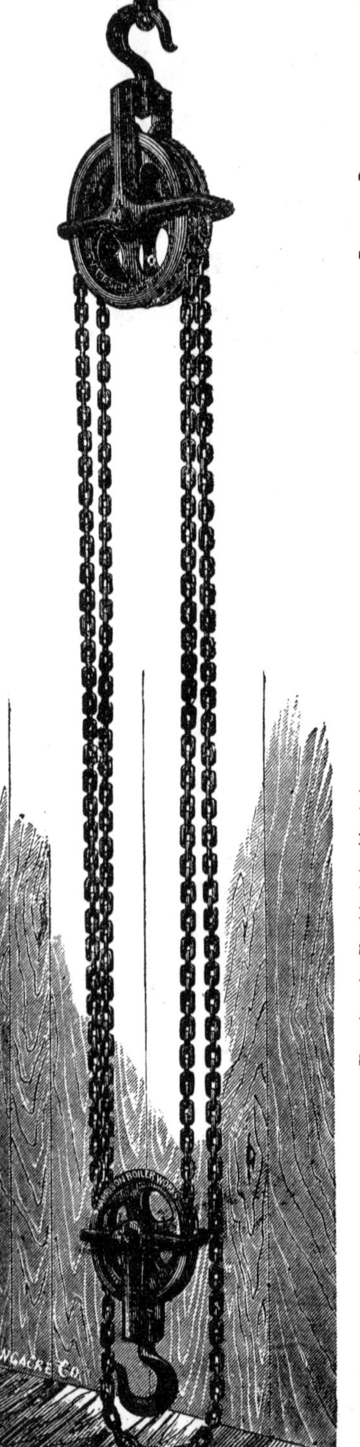

ONE TON PULLEY.

T. A. WESTON'S

PATENT DIFFERENTIAL

PULLEY BLOCKS

Manufactured under License from Patentee,

AT THE

Harrison Boiler Works,

GRAY'S FERRY ROAD,

PHILADELPHIA.

The peculiar merit attached to these Pulleys is, that while they are more powerful than ordinary pulley blocks, they also possess the novel and invaluable quality of not "running down" under any circumstances, while the load is suspended to them. Wherever weights have to be lifted, this hoisting tackle WILL BE FOUND MOST SERVICEABLE.

PRICE LIST.

Size of Block.	No. of Feet of Chain in each Block.	Price.	Price for Additional Chain Per Foot.
½ ton.	26	$25 00	40 cents.
1 "	30	30 00	45 "
2 "	38	50 00	55 "
5 "	30	200 00	90 "

The 5 Ton arranged with Gearing.

CAUTION.

WESTON'S

PATENT DIFFERENTIAL

PULLEY BLOCKS.

75,000 IN USE.

MEDALS:

World's Fair, 1862;

N. Y. State Fair, 1867;

Paris, 1867, etc.

Washington, D. C., June 8, 1867.

In an Interference between the claim of T. A. Weston and the patent of J. J. Doyle, the priority of Weston's well-known invention was fully established. The public are hereby cautioned not to buy or use any Differential Pulleys marked "Doyle," "Bird," or "Longley," legal proceedings being now in progress for the suppression of infringement. Parties, who have purchased infringements of Weston's patent, may obtain licenses legalizing the use of the same, upon reasonable terms, on application to

T. A. WESTON,

43 Chambers St., New York

Or his Attorneys,

ABBETT & FULLER,

229 Broadway, New York.

For Terms, etc., address

HARRISON BOILER WORKS,

PHILADELPHIA, PA.

FIVE·TONS PULLEY.

THE HARRISON BOILER.

Scale ¼ inch to the foot.

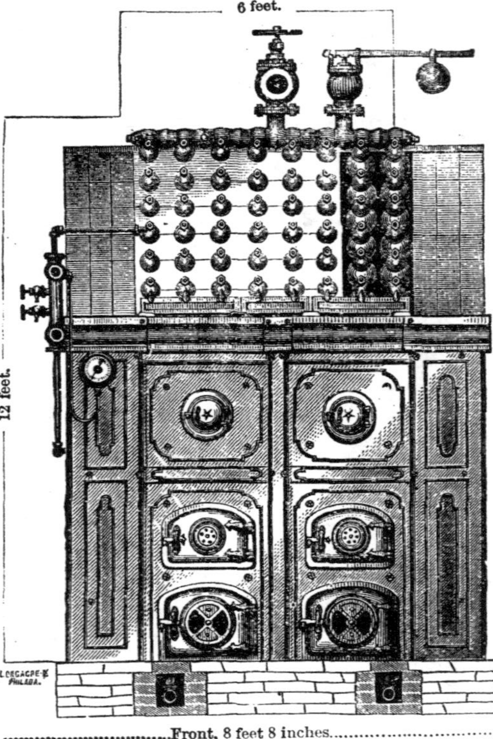

Front, 8 feet 8 inches.

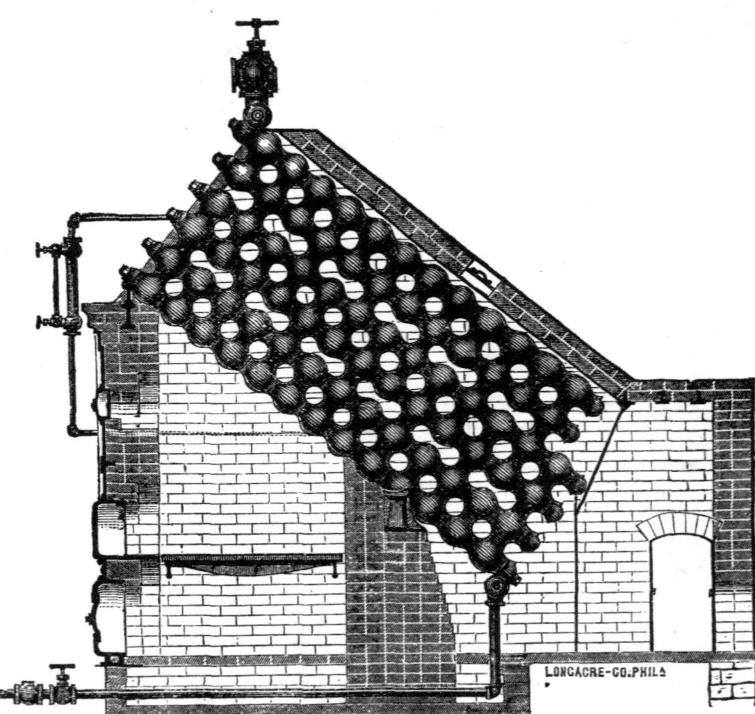

Side, 13 feet.

CANNOT BE BURST UNDER ANY PRACTICABLE STEAM PRESSURE.

CONTENTS.

FIRST-CLASS MEDAL, AT THE WORLD'S FAIR, LONDON, 1862.

"FOR ORIGINALITY OF DESIGN, AND GENERAL MERIT."

ONLY FIRST MEDAL AND DIPLOMA AT THE AMERICAN INSTITUTE FAIR

NEW YORK, 1869.

HARRISON BOILER WORKS,

PHILADELPHIA.

At the American Institute Fair, the Harrison Boiler received the only First Medal and Diploma over all others. There were seven on Exhibition, beside an ordinary Firebox Tubular Boiler.

Two Harrison Boilers of fifty horse-power each, were exhibited, and as will be seen by the official report, were the only boilers put in operation on the ground, that were found reliable, and capable at all times of doing the work they were built to perform.

During the test of the Corliss Steam Engine, built by W. A. Harris, of Providence, R. I., these boilers, with feed water at a temperature of 47 degrees, furnished a horse-power for $3\frac{19}{100}$ pounds of coal per hour; developing seventy-six $\frac{57}{100}$ horse-power of boiler, being within $23\frac{43}{100}$ per cent. of their capacity. During part of the fair, they developed one hundred and twelve horse-power.

The principal competition was with two sixty-horse power Root Wrought Iron Sectional Boilers, advertised in the New York papers as of that capacity, and being sold to perform that amount of work. Owing to the priming of these boilers, and the leaking of steam and water, at the gum joints, at both ends of the tubes, they were found incapable, after repeated trials, of driving the larger engines, and were used to supply steam to the pumps on exhibition.

The Harrison Boilers were used for supplying steam to the engines, of which there were two of eighty, and three of fifty horse-power, requiring a large volume even without load.

Enclosed, please find Official Report on Boilers; also, Certificates of Engineers, and Builders of Engines on Exhibition.

REPORT ON STEAM BOILERS.

Thirty-eighth Fair of the American Institute, held in the City of New York, October, 1869.

THE HARRISON SAFETY BOILER.—First medal and diploma for, 1st—Safety; 2d—Economy of Space; 3d—Economy of Fuel. This Boiler was the only one which was found reliable and capable of driving the engines at the Exhibition, and which did furnish all the steam for the competitive test of the engines.

ROOT'S WROUGHT IRON SECTIONAL BOILER.—Second medal and diploma for facility of repairs and economy of space.

A true copy from the report on file adopted Dec. 7th, 1869.

(Signed),

JNO. W. CHAMBERS, *Sec'y.*

FAIR OF AMERICAN INSTITUTE.

NEW YORK, October 30th, 1869.

We the undersigned hereby certify, that the Two (2) Fifty (50) Horse-Power Harrison Safety Boilers, have furnished the steam to the five Steam Engines on Exhibition during the greater part of the past five weeks.

The Steam has been of good quality—abundant in quantity, and of uniform pressure.

The Boilers are tight, and have shown no leakage during the Exhibition—and their performance has been entirely satisfactory.

The Harrison Boilers were used during the present week in testing the Engines on trial.

WM. H. BUTLER, JR., Engineer Exhibition,
S. W. BLOOMER, " Rider Engine,
W. L. CRAIG, " Corliss,
T. D. BLAKE, of Geo. F. Blake & Co., Steam Pump Manufacturers,
A. W. HARRIS, Harris Engine, (Corliss cut-off),
COLE BROS., Steam Fire Engine Builders,
PATRICK CLARK, Patentee of Clark Blower,
JOHN COLLINS, Engineer,
W. A. HARRIS, Builder of Corliss Engine, Providence, R. I.,
WM. BAXTER, Patentee of Baxter's Portable Steam Engine,
GEO. C. CREAGER, in charge of Merrick & Son Steam Hammer,
C. M. VAN LINE, Loomis Engine.

For Descriptive Circular, Pamphlet and Price List, address

HARRISON BOILER WORKS.
PHILADELPHIA.

"*All boilers should be so constructed that their explosions may not be dangerous.*"

DR. ALBAN,
Of Plau, in Mechlenburgh.

"Corrosion corresponds, in its comparative frequency and fatality, to the great destroyer of human life, consumption. It is the one great disease."

ZERAH COLBURN,
Before British Association, at Bath, 1864.

"Furrowing along a seam of rivets, or rather under the line of an overlap, is found to be the usual malady." "So far as furrowing is concerned, there can be no doubt that *wrought iron is the worst material* that can be employed for a steam-boiler."

*Report of Manchester and Midland
Boiler Association, for* 1863.

"*Steam-boilers can no more be made absolutely secure against some kind of explosion or fracture than guns or ordnance.* But they should be and can be made, so that no serious harm can arise when they do give way. To accomplish this most important end, the prevailing system has been found after a century of trial, entirely at fault, and improvements must be looked for in its abandonment."—*Public Ledger*, June 14th, 1867.

A verdict of the Coroner's jury, in the case of an explosion at an Iron Works in Kensington, April 27th, 1868, signed by some of the first experts in Philadelphia, says:—

First. "That the defect that occasioned the accident was not apparent, nor could it have been detected until the small leak occurred on the morning of the 27th of April."

Second. "That the immediate detection of the defect, and the prompt attention given to it, shows that care and the *intention of safe working*, on the part of the engineer and managing partner, had been given in this instance, and that the *error of judgment* which allowed the use of the boiler after it had shown its defective condition, was one which the jury deplore—*they cannot blame* those who committed it." This "intention of safe working," and this "error of judgment," caused the loss eventually of six valuable lives.

AN ESSAY

ON

THE STEAM BOILER.*

AMONG the elements that have been pressed into man's service, water turned into steam holds a most important place. And strange as it may appear to the uninformed, it might almost be said, that the steam-engine, as matured by James Watt, came from his hands nearly perfect in principle,—armed and ready to do battle, in the varied fields in which it has since been employed. James Watt knew all, and acted with a knowledge of all, or nearly all, the principles that are now known. Improvements in the steam-engine of our time, consist in a better arrangement and proportion of parts, better material, better workmanship, and vastly increased size. Many of its better qualities are the result of improved means of manufacture, which with equally improved quality of material, has enabled the steam-engine builder to do such work as could not have been done under a less improved system, and for which Watt might have sighed in vain.

Not so the steam-boiler. It, from the very first application of steam as a useful agent, has been the constant trouble of the engine-builder and the engine user, the great source of anxiety, danger and expense. The first patent regularly issued in England for a steam-boiler, dates about a century back, and from that time to this, patents for new designs or improvements, numbering thousands, have been issued in the United States,—in England, and on the continent of Europe. Notwithstanding the vast amount of labor and thought that has been bestowed upon the subject, the whole engineering profession still is in doubt as to which is the best steam-boiler, no single one, at this

* Read before the Franklin Institute, January 16th, 1867, by Joseph Harrison, Jr., Philadelphia. (Revised and in part re-written.)

moment, proving so much better than the legion that surrounds it, as to take any very prominent place in the general estimation, and not one combining the most important principle of security against destructive explosion.

Stone, wood, cast and wrought iron, copper, steel and various alloys of other metals, have been tortured, bent and twisted, from the beginning, into almost every conceivable form to make a steam-boiler. Still the work of change goes on, patent upon patent being continually issued for attempted improvements in this much needed object. It is remarkable that changes have tended more towards saving weight, cost or fuel, than in the more important object of making it safe from explosion.

The paramount aim in the use of steam, should be safety, and yet, with all that has been done, no steam-boiler now in general use *approaches* this essential requisite in its construction, compared with what is demanded of it. Hence the frightful loss of life—the dreadful maiming and suffering, the immense amount of valuable property annually destroyed by steam-boiler explosions.

It may be said that there is no remedy for this state of things,—that all has been done that skill and ingenuity can devise, to stop such fearful results. If we *have* arrived at the end, and found no remedy, then must we accept the situation, trusting rather to Providence, care or chance, to protect us from harm, than to any inherent controlling principle in the thing used, voting steam a good servant but a very bad master.

Before concluding this paper, I will endeavor to show that all has not been done in the general use of steam, to render it as safe an agent, as its wide-spread utility and necessity demand. Nay, more; it will be shown, from many years of practical experience in the use of a steam-boiler of singular design, and of material not heretofore considered best for the purpose, that the employment of steam at any practicably useful pressure, *can* be made, almost entirely harmless.

Some writers give to Dr. Alban, of Plau, in Mecklenburg, the credit of first enunciating the grand idea that "*all boilers should be so constructed that their explosions may not be dangerous;*" but it is scarcely possible that Evans, Hancock, Gurney and others at a much earlier date, should not have as fully appreciated this most important principle.

When the low-pressure of the earlier era of the steam-engine was used, the form or material of the steam-boiler mattered little, and we find Savery using cast-iron, Newcomen wrought iron; but from the difficulty of getting good plates of the latter material, Watt even made and put in use boilers of *wood,* hooped in the manner of the soap-boiler's kettle, with cast iron curb at the bottom. But when the first really high-pressure engine was introduced by our countryman, Oliver Evans, carrying steam as high as *one hundred pounds to the square inch,* and upwards, it then became necessary to look for material and form capable of sustaining such pressure. Oliver Evans used wrought iron plates in plain cylinders of small diameters, sometimes, with internal return flue, through which the heated products of combustion passed, after coursing the whole length of the lower half of the boiler.

These two kinds of boiler are at this day more extensively used in the United States than any other, and may be found almost exclusively on our Western river steamers. Perhaps no other boiler now in such general use, has greater safety in its principle of construction than this of Oliver Evans.

It is true that the most disastrous explosions on record have occurred with cylinder boilers on our Western rivers, but these calamities have been the result of scanty proportions in the first place, in order to save cost and weight, or from depreciation after long use, rather than from defect in principle. If the grand idea insisted upon by Dr. Alban be the true one, then have our engine-builders wandered far away from it since the days of Oliver Evans. Look at the immense structures built up of wrought iron, now so largely made for, and used on ocean and river steamers! Is this principle of safety attained, or even aimed at, in these boilers? Are they so made that "*explosions are not dangerous?*" Witness the disaster on the North River steamer *St. John,* in 1865. Here a boiler exploded, made on an approved and often used plan, which, according to the testimony of experts on the Coroner's jury, "*pulsated*" at every stroke of the engine. Has any one seriously considered what this "*pulsating*" means? If anything, it means a movement in certain parts which, being kept up for a given and almost calculable length of time, must inevitably destroy the structure of the material of which these parts are made. It is not very assuring to the traveling public, that all of the best ocean steamers, as well as those on our rivers, lakes and sounds, have at this moment, boilers theoretically, if not actually, as unsafe as the one that blew up in the *St. John.* It is not too much to say, that all boilers of large dimensions, whether of square form, dependent upon stays or braces for their strength, or cylinders of large diameter, with or without internal flues, cannot be safe. Neither is it too much to say, that no boiler is safe, that can, under any circumstances, *rend and scatter large masses of material, liberating at the same time large volumes of highly charged water and steam.*

Take a boiler, if you please, that depends entirely for its strength upon being properly stayed, and there are thousands of such in use, especially for marine purposes. In the nature of boiler-work under such a system of construction, with the very best skill and the very best oversight, it is impossible to execute the work so as to be invariably reliable. Unlike almost every other class of work, even after the parts are finished, no matter how close the inspection, it is not possible always to know whether it has been well done or not. Let any one, with a full knowledge of the subject, watch the making of such a boiler. The drawings are perfect, every strain calculated to a decimal, every proportion exact. If it were possible to execute the work just as laid down, all might be well; but if such a thing is possible, we have never seen boiler-work made with such accuracy. In the matter of the stays (a most important point), every hole should be exactly smooth and true, and made to come in true line with the one it has to meet. Every bolt should be turned and fitted to its appropriate hole. But all who are acquainted with boiler-work, know that it is not even attempted to do it in this manner. Ill-shaped stays, badly made and badly fitted, or strained into ill-shaped places, often out of reach of the eye and the hand of the workmen;—rough holes most frequently made in the smith's shop, with as roughly made bolts. If the holes are bored, so rudely do they usually adjust themselves to one another, that the ever-ready drift—that bane of safe and good boiler-work—is resorted to, bringing the work together under a tension that puts to flight all decimal calculations, and but too frequently dismembers the parts themselves. A boiler, when finished, may be submitted to a water-pressure test; but this must not be too much relied upon, as this very test may so strain a steam-boiler, that a much less steam-pressure when fired, might lead to disaster. Can such a boiler, dependent upon such work, be safe? And again, take plate-riveting. An English writer says:

"It is a truism, 'that the strength of any structure is its weakest point; but who can say where the weakest point of a steam-boiler is, as ordinarily made?'" "Take a simple cylinder boiler, for instance,—the sheets are run through the rolls and bent to the proper radius, and when the riveting gang get to work they close up the rivets with great rapidity; but when the holes come out of line with each other, the drift-pin is resorted to, and the sheets are literally stretched until the rivets can be inserted; when the drift-pin is knocked out, the sheet goes back to its place, and there is already, without a pound of steam-pressure, strain enough to cut the rivets off." "Repeat this performance through twenty or thirty feet, the length of an ordinary cylinder boiler, and who can say where the weakest point of the structure is? Suppose such a boiler made of silk or any flexible material, what shape would it be in?" "It would be full of puckers, folds, seams and gathers, and represent most accurately the various trials to which that most abused of all modern engineering apparatus—the boiler—is exposed." "The case is aggravated, not benefited, when we construct a square boiler, for this shape seems, by general consent, to have been adopted for marine service."

"When the angles or flanges of the sheets are not broken by the flange-turners, they are cracked out by the drift-pin of the riveting gang, and it ought to be made a capital offence to have such a tool on the premises of any boiler-works." "New boilers burst under the most mysterious circumstances; old boilers are patched and then burst; and we are told that 'putting new cloth into old garments is the solution of the trouble.'" "On each occasion the Coroner examines a host of 'experts,' who proceed to declare that the 'iron was burnt,'—'the water low,'—'the stays insufficient,'—'the water changed into explosive gases,' etc.; but it never occurs to these worthies, that the actual strength of the boiler was, in many cases, unknown, and that it may have been at the bursting point for many days, weeks, or months, until at length it gave way." "It is ridiculous to suppose that safety is secured by neat-looking rivet-heads or handsomely caulked seams." "Holes will come out of truth with the utmost care, especially in such hap-hazard work as punching." "Neither are the braces (stays) properly set; for some draw all one way, while others do not draw or hold at all, and are perfectly loose. Thus a portion do all the work, and the rest are idle;—they impart no strength, and are an element of weakness; for the engineer relies upon them when they are doing no good." "We are confident that a great deal of attention can profitably be given to the mere workmanship of steam-boilers;—they are not tanks for boiling water, but great magazines wherein tremendous power is stored, the safe custody of which is of paramount importance to all in the vicinity."

Assuming that a boiler of large dimensions, whether cylinder or marine, *can* be made so that all the parts are joined together without strain, this state of things can only exist at the uniform temperature throughout, under which the boiler has been made. Put fires at white heat into or under such a boiler, heating the plates in the immediate vicinity of the fire, as must occur, in a much greater degree than at the external or more remote parts of the structure: surely then the parts that had previously lain quietly together, assume a new and constantly changing condition, and who can tell what these changes are, their frequency, or to what extent the strength of the boiler is impaired thereby?

Let us now turn our attention to another equally, or perhaps more, important point than those we have been considering,—the wear and tear of plate-iron boilers. A writer in the *London Mechanics' Magazine* says: "It is not too much to say, that nine out of ten explosions are directly the result of corrosion." "Setting aside the value of human life and limb, we find that the mere pecuniary interests involved in either the gradual or sudden destruction of a boiler, are very considerable." "Repairs are, at all times, expensive, and the time lost in making them is often a serious source of pecuniary loss, worry and trouble." "Hence the replacement of a plate, or the alteration in a defective flue, is often staved off from day to day, until irreparable mischief is done." "Reflecting upon these things, it seems strange that boilers are made, fired and worked with a negligence, which apparently regards iron plates as indestructible, and the results of an explosion trifling to a degree." "We cannot set such a system, or rather such a want of system, down wholly to stupidity or neglect." "We know that boilers in the best hands, and under the most careful management, often become worthless with a startling rapidity, which no amount of theoretical reasoning can account for, nor practical skill arrest or delay." "The utter uncertainty in which the engineer is doomed to live, as to what does or does not promote durability, leads naturally to recklessness, neither the result of the want of thought nor indolence." "Corrosion is too often regarded in the light of a fate—a destroyer, merciless and indiscriminate, before which, as a *fetish*, the manufacturer and ship-owner bow down and submit."

Mr. Colburn, in a paper read before the British Association, in 1864, says: "As a boiler malady, corrosion corresponds in its comparative frequency and fatality to the great destroyer of human life, consumption. It is the one great disease." "A trickling of condensed steam down the outside of a boiler will inevitably produce corrosion, and to this was directly traced a large number of the forty-seven boiler explosions which occurred in the United Kingdom in 1863, and which caused the loss of seventy-six lives, with injuries, more or less serious, to eighty persons."

In the report of the Manchester and Midland Boiler Association, for 1863, we find the following: "Furrowing along a seam of rivets, or rather under the line of an overlap, is found to be the usual malady, but the iron is eaten away almost everywhere; not uniformly over the whole surface, but in numberless holes." "So far as furrowing is concerned, there can be no doubt that wrought iron is the *worst* material that can be employed for a steam-boiler."

Thus much on the subject of corrosion. Another article in the *London Mechanics' Magazine* says: "Until a comparatively recent date, the belief obtained with most engineers, that a riveted joint, if the work were properly done, was superior to the plate itself."

Mr. Wm. Fairbairn, in a series of carefully conducted experiments, upset this fallacy by proving, that, "the strength of the plate being taken as one hundred, that of a double-riveted joint will be seventy, and a single-riveted joint fifty-six;" and this with first-rate workmanship. "Fifty-six per cent. of the whole strength of boiler-plates is certainly not much to realize with the best workmanship; but as many boilers are put together, this percentage must be regarded as too high." "There are difficulties involved in the nature of the process, which the best mechanic can only combat—seldom or never overcome." "However accurately two plates may correspond before being punched, that process inevitably distorts them, and occasions a bad fit when subsequently put together." "The hammering and bending at the edges is invariably injurious to cold plates."

"Again, the best workmen, with the best machinery, find it out of the question to make all the holes in a long seam correspond." "The constant use of the drift is certain to follow, and when plates are of inferior quality or very thin, cracks are frequently established from one hole to the other."

"The judicious use of the caulking chisel easily conceals the defect, which is none the less serious because it is invisible." "The best rivets too seldom completely fill the holes they occupy." "They are never truly at right-angles to the plates, and are often exposed to enormous strain in drawing plates together when they are badly fitted." "We have seen, from this cause, the heads fly off half a score of '*Best, Best*' rivets at once, in rolling a new boiler from one side of the shed to the other."

Blistering of plates is another trouble in the use of plate iron. *Engineering Facts and Figures*, for 1863, page 21, says: "The fact of plates by good makers being liable to blister unawares, and which previous examination fails to detect, shows the importance of not hazarding an expression upon their soundness. Thus the strength of no unassisted plate, exposed to the action of the fire, should be relied on, and consequently it becomes most desirable that furnaces should be in every instance stayed either with flanged seams, or with hoops of angle iron, T-iron or other advantageous form."

Thus at every turn, the boiler-maker, in using wrought iron, meets with difficulties which can only be partially, never perfectly, overcome. These difficulties occur most frequently at the very points in the structure where danger from defective work or material is most imminent, and where it is least easy to avoid it.

The maintenance of a well-made steam-engine is of slight importance. So true is this, that engines are doing good service now in England, that were made by Watt and his contemporaries, the sun and planet-wheel even yet making their regular revolutions. Where are the boilers that started with these engines? Gone, gone, and many succeeding the first gone also.

The elements that destroy a steam-boiler commence their work from the moment of its completion, and from the hour it is first filled with water and fired. Whether used or not, the insidious process goes on, and it is fortunate if the life of the boiler reaches a decade, ere it is thrown out as worthless. From *Engineering Facts and Figures*, for 1865, we quote the following:

"The saying of that distinguished authority in matters mechanical, —Wm. Fairbairn,—'that danger in the use of high-pressure does not consist in the intensity of the pressure to which the steam is to be raised, but in the character and construction of the vessel which contains the dangerous element,' may be set down as a truism, containing a great deal of suggestive truth, but which is often overlooked, if not entirely ignored."

"Else how is the public sense of what ought to be, but unfortunately is not, every now and then shocked by a recurrence of those accidents which result in such extensive loss of life and property." "It is the saying of one who has said many good things in his day, that 'self-interest is always intelligent.'"

"In the matter of the use of boilers notoriously defective in form, material and construction, self-interest is *not* always intelligent; for however easily employers may take the loss of life from accidents in the use of steam-boilers, one would think that self-interest would prompt them to avoid, by all means in their power, the loss of property."

What are the conclusions that are forced upon us by all that has been adduced? Plainly that wrought iron has proved itself unfit for steam-boilers; that it is unreliable and unsafe to use it for such purpose, and that neither in principle nor workmanship, in the use of this material, have we advanced one step, in a century, towards making the steam-boiler, as now generally used, safe from destructive explosion. On the contrary, just in proportion as we have increased the working pressure, so have we run into greater danger; and at this moment boiler explosions are more frequent, and more fatal in their consequences, than ever.

It is a sad condition of things that this much needed force, should be so little within control. Must these mines of destruction, placed in our cities and towns, under our feet as we tread the sidewalk, and all around us, still hold their pent-up wrath by so frail a thread? Is there no way to render them more safe?* I think there is a way to do

---

*On September 9th, 1867, an explosion occurred in West Twenty-eighth Street, New York, in which a steam-boiler, weighing, as stated at the time, nearly five tons, was projected high in the air, and descending at a distance, horizontally, of not less than four hundred feet from the place it had previously occupied, crushed its way through, and almost destroyed, a large dwelling-house in its course. It may be stated that the establishment in which this boiler was used, became, at the first instant of the disaster, almost a complete wreck, killing and burying two of the workmen in its ruins. Its sudden and fatal effect in the home of innocent and unsuspecting childhood, is thus described in the New York *Times* of September 10th, 1867:

"A visit to the dwelling of Mr. Hausman, disclosed another scene of destruction. The flying boiler struck fairly upon the roof, a few inches from the rear wall, with the upper and smaller end descending. So great was the force of the descent of the huge mass, that the rear of the building crumbled and gave way beneath the descending weight of iron, which crushed through all of the four floors, and finally landed in the cellar. At the time of the explosion, one of the female servants employed by Mr. Hausman, named Mary Dowling, was on the attic floor in the rear, and immediately below her, on the third floor, were Maria Weiberzahl, the wet-nurse, and the infant, Henry Hausman, aged eleven weeks. The child had just fallen asleep, and the nurse had placed him in the crib while she began dressing herself. In the bath room, adjoining the nursery, was another child, Dora Hausman, aged nine years, who was bathing. The descending boiler fell on this portion of the building, and enveloped all four of these persons in the *debris*, the two children being instantly killed, the wet-nurse being severely injured, and the domestic, Mary Dowling, sustaining fatal injuries."

this. If this can be shown, then let no one say hereafter that steam-boiler explosions cannot be prevented.

Mr. Wm. Fairbairn, whom we again quote, says: "Instead of working two hundred pounds pressure to the square inch, I think we shall reach five hundred pounds."

In *Engineering Facts and Figures*, for 1863, in treating of the great need of improvement in marine boilers, we find the following: "The answer is obvious,—no further economy can be obtained in steam-power without the use of *high-pressure* and expansion."

Ocean steamers, twenty-five years ago, used three or four pounds pressure to the square inch. Now the Cunard steamers use twelve or fifteen.

Our North River and Sound steamers, the pioneers in using much higher pressure in condensing engines, carry thirty or forty, and even fifty pounds to the square inch. Common consent, if not necessity, demands higher pressure, and it behooves the engineering profession to look to it, that we do not continue the present imperfect and most dangerous system, if there is any way to avoid it. Enough, we think, has been said to convince the most prejudiced that *a safe steam-boiler has not yet been made of wrought iron.* Assuming this to be proved, in what direction must we then look to find a better material for the purpose,—one, not possessing the defects of wrought iron,—one, that can readily be made into such forms as will most conduce to the safety, durability and economy of a steam-boiler?

Turn we now to cast iron,—early used, but heretofore and even now generally supposed inferior material for steam-boilers. A writer in the *London Mechanics' Magazine*, for May 2d, 1864, uses the following language:

"There is a French proverb which says, that we always return to our first love, and it is by no means unlikely that this will be verified in boiler engineering. At one period, it is beyond question that cast iron boilers were habitually used for very high pressures, and they were used because the material possessed constructive advantages which were not then believed to reside in wrought iron; and if these advantages reside in it still, under a principle of construction modified to meet existing demands, there is no good reason why it should not be habitually employed. Cast iron is far better adapted to meet the ordeal of fire and water to which a boiler is exposed, than the best wrought iron plates ever manufactured."

"As to strength, we all know, or ought to know, that that is a matter of proportion quite as much as a matter of material. There is nothing like practical illustration to bring such truths home to the mind. Let us suppose, then the case of two boilers, one made of plates half an inch thick, and the other one quarter of an inch thick. If each of these boilers is, say, six feet in diameter, the first one will possess, as nearly as may be, double the strength of the other. To render both of equal strength, it is only necessary to reduce the diameter of the thinnest one to half the diameter of the thickest."

"In the same way, it is certain that a cast iron tube, of a given diameter, may be made quite as strong as one of wrought iron of the same thickness, provided the diameters are proportioned, the one to the other, in the ratio of their tensile strength. That the arguments adduced against the use of cast iron are many and powerful, we do not pretend to deny; but that they are invariably applicable, or that it is, in other words, impossible to devise a boiler that shall elude these objections, is false."

"We daily see cast iron used to carry enormous pressures with the utmost confidence. Its tensile strength may always be brought, in one sense, up to wrought iron by using enough of it. It has thus beaten wrought iron, in the form of guns, many times."

"There are two ways of increasing the strength of any vessel; the one in increasing the thickness, the other in reducing the diameter of the globe or cylinder to be tested. It is obvious that cast iron can only be used in small tubes or chambers, inasmuch as large vessels must necessarily be of such a thickness, that heat would pass through it very slowly indeed. But this fact in no way militates against the safety, economy or efficiency of a generator. Perhaps the present system of employing wrought iron boilers of colossal dimensions in our every-day practice, has been productive of more injury to life and property, than can be laid at the door of the engineer on any other ground."

In a leading article in the *Engineer*, for 1864, it is said: "It has been so long the custom to consider cast iron as a brittle material, hardly to be trusted under pressure, that it requires some amount of reflection to perceive wherein it possesses manifest advantages over wrought iron. The resisting strength of a properly made cast iron boiler is calculable, and a good *a priori* case could have been made out in its favor long ago."

In an article in the *American Artisan*, for November 22d, 1865, in answer to an assertion made in that journal, referring to the Harrison

Boiler, that "cast iron was not to be recommended for steam-boilers," because it "was liable to be strained from inequality of temperature," I have said, "Many years of experience in the use of this boiler has taught me that, as a material for steam-boilers, cast iron is preferable to wrought iron, and for a reason that can be easily understood. Cast iron is *not* LIABLE TO BE STRAINED 'by inequality of temperature;' *it is liable to break* from such cause, and will give out at once if badly proportioned or improperly used. Wrought iron in steam-boilers *is liable to be strained by* 'inequality of temperature,' and not fracturing at once, goes on straining until its structure is destroyed, and the parts thus strained inevitably give way."

"Put cast iron in such a form as will prevent harm in case of rupture, and it becomes the *best* material for steam-boilers, and one of its best qualities is in giving out when badly treated. Not so wrought iron, its very tenacity begetting a false security, which might lead to disaster at any moment."

It certainly appears strange at a first glance, that brittleness in any material should make it more reliable than a more tenacious one, for purposes needing strength.

In the finding of the Coroner's jury in a recent boiler explosion in the city of Philadelphia, by which five men lost their lives, we find the following:

"That those in charge of these boilers exhibited culpable negligence in regard to the precautions universally considered as essential to the safe management of steam-generators, and were not sufficiently experienced to render their management of such apparatus safe. A similar want of knowledge, experience and care is only too common among those using steam-power."

"That the proprietor was guilty of neglect in failing to provide a competent engineer to take charge of the steam motors of the establishment."

"In connection with this occurrence, this jury reiterates what has been already most strongly expressed under like circumstances, that official inspection of all steam-boilers should be provided by our local government, and is as essential to the safety of life and limb among our citizens as any part of our police regulations. The storage of gunpowder in our city is prohibited by law; but any one may place a steam-power magazine, with match burning, at our side or under our feet, with perfect impunity." "Such magazines undermine, in fact, our most crowded streets, and, unless properly cared for, will one day reveal their existence in terrible disaster."

There is much for reflection in these extracts, and their teachings should not be passed unheeded. In the use of steam in stationary engines, of small size in our large cities and towns (and from these as much harm can arise as from larger ones), it is almost the rule to employ men of little experience or skill, who are utterly incompetent to fill the most important post of fireman.

The fact that they themselves are in the greatest danger, fails to make them more thoughtful, and they ignorantly and carelessly go on, from day to day, hanging on the very verge of disaster. Almost unaware of the tremendous power that is in their keeping, and failing to profit by the plainest evidences of danger, a long continued immunity often leads them recklessly to think that no harm can arise, even when danger is most imminent.

And in the matter of inspection of boilers of stationary steam-engines, insisted upon so strongly, something also may be said. It is no doubt true that the regular government inspection of steamboat boïlers has been productive of good; but, at the same time, too much reliance should not be placed upon even this safeguard. It is well known that a steam-boiler may pass inspection, and be, at the moment, all right; but it is equally well known that even in an hour thereafter, it may, by neglect, become entirely unsafe.

A steam-boiler may pass the most critical examination, and may stand a severe hydraulic test when cold, but a much less pressure, superadded to the strains resulting from irregular expansion, when hot, might rend the same boiler into fragments.

It has been conceded "that the strength of any structure is its weakest point." Let us now consider how a parity of reasoning will apply to the use of steam. In almost every case of boiler explosions, we have the changes rung upon the stereotyped words, "culpable neglect in some essential particular," "gross ignorance," or "carelessness," and that closer and more frequent official inspection, greater skill and more careful oversight on the part of those in charge, would have prevented the catastrophe. So it might, in some instances, no doubt; but it should be borne in mind, that neglect and carelessness (whether wilful or not), as well as the most stupid ignorance, may always be counted upon.

Hence we should adopt as an axiom, that safety is to be looked for amidst the greatest amount of ignorance, or even wilful carelessness and neglect, with which steam-generating apparatus may be used, without causing disaster, rather than in the exercise of the greatest care and skill.

It is worth while to reproduce here in part a leading article from the *Public Ledger* of June 13th, 1867, just one week after the dreadful calamity in Sansom Street, Philadelphia, when nearly thirty persons were hurried into eternity, maiming many others, and utterly destroying a large manufactory in a moment of time:

"PREVENTION OF BOILER EXPLOSIONS.

"The recent steam-boiler explosion warns us again, in tones not soon to be forgotten, that greater security to life and property in the use of steam has become a paramount necessity; and, leaving out all other considerations, the subject demands the earnest attention of every one who values his own personal safety. The public will be again told that ignorance and want of experience,—reckless, if not wilful neglect of the most obvious precautions,—material and workmanship of bad quality,—low water,—over pressure,—explosive gases,—electricity, or some other such thing has been the cause of this terrible calamity. But it must not be overlooked that steam-boiler explosions, largely destructive to life and property, have occurred more than once in the largest, most experienced and most carefully managed engineering and manufacturing establishments in our city.

"And it must also be borne in mind that it is quite impossible to procure boiler-plates of wrought iron invariably reliable, or that will remain so, no matter how much care is bestowed upon their manufacture or inspection. It is equally certain that the best material ever put together with the very best workmanship, under the present system, may, and does frequently, share the same fate as the most inferior. Thus, whether these much used appliances are made as near perfection as may be, directed by intelligent and skillful management, or the reverse, their end is but too often the same. Under such a condition of things it is not worth while to waste time in theorizing as to questionable causes. It is with the disastrous results and their possible remedies, that wise men should deal.

"Steam-boilers can no more be absolutely secure against some kind of explosion or fracture, than guns or ordnance. But they should be and can be made so that no serious harm can arise when they do give way. To accomplish this most important end, the prevailing system has been found, after a century of trial, entirely at fault, and improvements must be looked for in its abandonment.

"Councils, if they have the power, or if they have not, then the State Legislature, in their efforts to enforce stringent rules for the inspection of all boilers of the kind that are known to be liable to disastrous explosions, should try the experiment of an ordinance or law, that no boilers, after a certain date, shall be erected within our city limits, unless a certificate is first obtained from competent inspectors, that an explosion of said boiler *will not be dangerous.* There is no full security in any direction, pass what law you will, make what inspection you may.

"The time is not far distant when the present system must come to an end, unless it can be proved beyond all doubt that no change for the better is possible. The pnblic safety now demands that these engines of destruction shall no longer be kept hidden from sight in the basements of our most densely peopled neighborhoods, or in our crowded workshops, and even under the busy footways over which we unconsciously tread, always ready for havoc, and only perhaps temporarily held from repeating the horrors of last Thursday, by a constantly weakening chain, the real strength of which is seldom if ever known.

"If these destructive engines were new things, and it was now first proposed to place hundreds of them in our crowded thoroughfares and under our pavements, who would listen to such a proposition? It would not be entertained for a moment. Neither should the system itself, although it is well established, be tolerated for a day beyond the time when it can be dispensed with."

It is beyond controversy or doubt, that the fracture of steam-generating apparatus, can and may occur, under any system of construction. No amount of official inspection or special care can entirely prevent this, and it must be set down as an imperative law, if safety is to be secured, "*that all boilers should be so constructed that their explosions may not be dangerous.*"

Having said thus much upon the general question of steam-boilers, it now becomes my province to describe the

HARRISON STEAM-BOILER,

An invention of my own, now being largely made and rapidly introduced into practical use. I had long turned my attention to the subject of improving the steam-boiler, and believing that better guiding principles were needed, I at length fixed in my mind the following axioms:

1st. That a steam-generator of whatever form or material, must, as a paramount condition, be absolutely secure from *destructive explosions, even when carelessly used.*

2d. That it must be constructed upon a system or series of uniform parts, simple in form, few in number, easily made, and easily put together or taken apart, and not of costly material.

3d. That its strength should in no respect be dependent upon any system of stays or braces, whereby the inefficiency or rupture of one of these braces or stays could cause greatly increased strain upon the others, thus endangering the whole structure.

4th. That its parts should not be of great weight or size, thus permitting greater portability and greater facility for getting it into or out of place.

5th. That it must have a principle of renewal, allowing the easy displacement and replacement or interchange of any one or more of its parts, without disturbing or impairing the material or workmanship of the remaining portions of the structure.

6th. That a boiler, whether of large or small dimensions, should have uniformly such elements of strength, as would render it always capable of safely sustaining many times greater pressure, than need ever be demanded of it in practice, and that its safety should not be impaired by corrosion, or the many other harmful influences, which so soon and so seriously affect the strength of ordinary boilers.

7th. That the parts should be so made and put together, that in case of rupture of any portion of the boiler, no general break up of the structure could occur, the release of the pressure by such rupture merely causing a discharge of the contents, without explosion or serious disturbance of any kind.

8th. That it should be constructed so as to facilitate the certain removal of deposit from its interior or exterior surface.

These axioms being carried out, all else being equal, it is believed that the result must be a better and safer steam-generator than any that has preceded it. It is not important here to enter into the detail of how the "Harrison Boiler" reached its present form. Suffice it to say, that when the form of a hollow sphere and curved neck was decided upon, as shown in the illustrations herein, it seemed that the true principle had been arrived at—that nothing more could be desired in that direction.

HARRISON'S CAST IRON STEAM BOILER.

Detail of 4-Ball Unit.

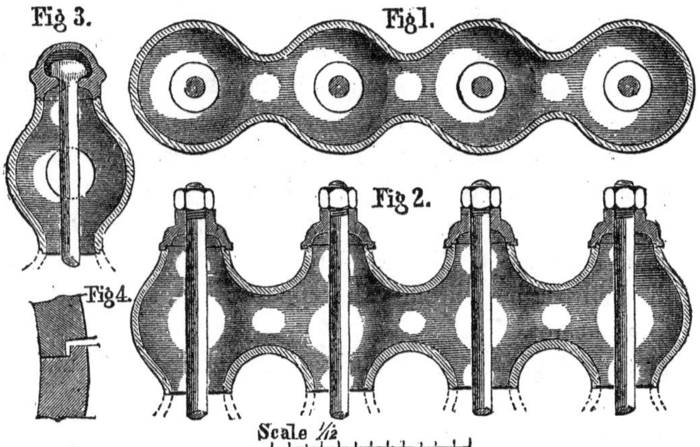

Fig. 1. Transverse Section.
Fig. 2. Longitudinal Section.
Fig. 3. Sectional Plan.
Fig. 4. Section of Joint. Half Size.

A boiler of about seventy-five horse-power, the first of the kind, was made of these simple elements, in the spring of 1859, and put in operation at the establishment of Messrs. Wm. Sellers & Co., Philadelphia.

This boiler effectively and economically supplied steam for driving the extensive workshops of the above firm, for several months. Its trial settled the question previously in doubt even with myself, that a boiler, as hereafter to be described, could be made and used, without its integrity being disturbed by irregularity of expansion and contraction consequent upon the action of fire.

In practice up to the present time, it has been usual to cast the spheres of eight inches in external diameter and scant three-eighths of an inch thick, placed in groups of two and four, making what may be called a brick and a half brick. These spheres are arranged in a straight line, one inch apart between their external diameters, and are connected with each other by a curved neck about three and one-quarter inches in diameter inside, at the smallest part. A series of half-necks make openings entirely through each of the spheres, at right-angles to the necks previously described.

These groups of spheres are called "units," and when jointed with a rebate-joint, accurately made on the edge or least diameter of the half-necks, fit closely together, making, when these edges are adjusted and drawn together, a steam and water-tight joint. Any number of these

3

two and four sphere units, when placed together with break-joints, may be conveniently made into a parallelogram or other form that may be desired. The distance between the centres of the spheres in the direction of the jointed or half-necks, when the units are laid together, is the same as the distance in the transverse direction, thus making an uniform section of any given length or breadth. Wrought iron tie-bolts pass through each line of spheres in the direction of the half, or jointed necks, connecting at each end with caps that close up the external orifices of the section. One of the caps, called a blank cap, is made to receive a T-head at one end of each bolt; the other end of the bolt passing through the opposite cap, and ending outside with screw and nut.

It will be seen that these tie-bolts will bind all the units in one section firmly together.

A section of six spheres wide, the upper rows twelve, and the four lower rows thirteen spheres long, will make a boiler of about six horse-power.

In setting a boiler ordinarily for stationary purposes, one or more of the sections are placed on edge, at an angle of about forty degrees, side by side, usually one inch apart between the sections, the upper corner on the bottom line being supported on a cast iron rail or bearer, the lower portion resting on a chair, adapted to the purpose, placed behind the bridge wall. A common steam-pipe, taking steam from the upper caps, connects any number of sections together. A similar pipe, in like manner placed at the bottom corner of the section, makes a water-connection between any number of sections.

To avoid binding of the parts from irregular expansion or disturbance of any kind, the steam and water-pipes are made up of short pieces, joining with spherical joints, held together with tie bolts, after the manner of the units. By this means a flexible connection is made, that prevents all trouble in joining the sections together.

No attempt is made to provide steam room, other than the capacity of the spheres in the upper angle of the sections above the water-line.

It is found in ordinary practice with the largest boilers, using high-pressure steam through an engine, that the steam-room allowed as above is quite sufficient. A larger proportion may be required for low-pressure engines, or where steam is taken out at irregular intervals in large quantities, and this may be easily had by adding to the number of the upper units, or by the use of a steam-drum.

The heated products of combustion rise from the grate, passing between the sections, and amongst the spheres, and over the bridge-wall, finding exit at length towards the chimney, at the lowest and coldest part of the boiler. A cast iron guard is inserted between each section, nearly horizontal, a short distance below the water-line. This guard prevents active heat from reaching the steam-spheres, but sufficient heat reaches the upper angle of the boiler, to dry and superheat the steam, ere it reaches the outlet to the steam-pipe at the upper cap.

It is not essential that a stationary boiler should be set exactly as described, as the units may be built up vertically, horizontally, or in any irregular manner best suited to the circumstances of the case. A stationary boiler set after the manner first indicated, and heretofore most generally adopted, has given very good and very economical results, when compared with other boilers having the best reputation. A boiler for marine purposes may be made in the form of a cube, and may be set without brick-work.

One of 160 horse-power has been in successful operation for several months in a steamer running on Chesapeake Bay from the port of Baltimore.

Having, as I think, in the previous pages, fairly proved that cast iron may be preferred to wrought iron as a material for steam-boilers, let me say a few words in regard to the advantages of the former material as used in the "Harrison Boiler."

Mr. Zerah Colburn, late editor of the London *Engineer*, and now editor of the new magazine published in London, called *Engineering*, has written much and very ably on the steam-boiler. His opinion, therefore, is entitled to great respect.

In a paper on steam-boilers, read by him before the Institution of Mechanical Engineers, at Birmingham, on May 5th, 1864, in alluding to mine, he says:

"Although it cannot be said that cast iron is in itself a strong material for boilers, yet it will be seen that in the form now described, it affords greater absolute strength against bursting than is possessed by any form of plate iron boiler. In a 'unit' of four spheres, each sphere having an internal diameter of about seven and a quarter inches, the whole area of the plane in which a bursting pressure could act, taken through the eight openings of the four spheres, is two hundred and twenty square inches, whilst the least section of iron resisting this pressure in the same plane is twenty-seven and a half square inches. In tensile strength may be safely taken at five and a half tons to the square inch. At this rate the bursting strength of the units would be one thousand five hundred and forty pounds."

And again, Mr. Colburn, in a paper read before the British Association for the Advancement of Science, at Bath, in 1864, says: "In ordinary boiler-making the geometrical advantage of the hollow sphere cannot be turned to account. It cannot be produced economically in plate iron, nor if in plate iron, could it be advantageously employed in a steam-boiler.

"The hollow sphere has this property, to wit: with a given thickness of metal it has twice the strength of a hollow cylinder of the same diameter. This is upon the assumption (which is correct when the cylinder is of a length greater than its own diameter) that the ends of the cylinder offer no resistance to a bursting pressure exerted against its circumference.

"Under over-pressure, a closed cylinder would take the shape of a barrel, and if of homogeneous material and structure, it would burst at the middle of its length in the direction of its circumference. The circumference of a sphere of a diameter of one, being 3.14159, the sum of the length of the two sides of a cylinder of the same diameter, and having a plane of rupture of the same area, is 1.5708, or exactly half as much."

And in the same paper, he says: "The tensile strength of cast iron varies between five tons and fifteen tons per square inch. Considered as a material for boilers, only the minimum strength should be regarded." "Cast iron boilers of eight feet in diameter, and of great length, were at one time made, but these were manifestly objectionable. The spherical form of a moderate diameter is preferable, and whatever is the strength of a riveted wrought iron cylinder, that of a cast iron sphere of the same diameter and same thickness of metal, will be the same.

"Plate iron of a strength of eighteen tons per square inch is virtually weakened to ten tons by the loss in riveting, and as the hollow sphere is twice stronger than the hollow cylinder of the same diameter and thickness, the cast iron having no joints, becomes equal in this comparison to the wrought plates."

"If we could always count upon the maximum strength of iron, to wit: twenty-seven tons per square inch for wrought and fifteen tons for cast, a fourteen feet cast iron sphere would have the same strength to resist bursting as the seven feet cylinder of the Lancashire forty-horse boiler, supposing the same thickness of metal in each case."

"But there is no occasion to make a boiler as a single large sphere; for it is now ascertained from extensive experience that hollow cast iron spheres of small diameter do not retain the solid matter deposited by the water. Small water-tubes, and indeed all small water-spaces in ordinary boilers, always choke with deposit when the feed-water contains lime; but cast iron boiler spheres, although they may be temporarily coated internally with scale, are found to part with this whenever they are emptied of water. *This fact is the most striking discovery that has been made in boiler engineering.*" It removes the fatal defect of small subdivided water-spaces, which can now be employed with the certainty of their remaining constantly clear of deposit.

"This discovery has been made in the use of the cast iron boiler invented by Mr. Harrison, of Philadelphia, United States."

"In Manchester, with feed-water taken from the Irwell, or from the canal, a hard scale is soon formed in the ordinary boilers; but in the cast iron boiler, a succession of thin scales of extreme hardness are found to form upon, and to become detached of themselves, from the inner surfaces of the water spheres. The scales are blown out with the water at the end of the week, and only small quantities can be found when purposely sought for. A pint of loose scales and dirt is the most that has yet been found in a careful internal examination, after nine months daily work." "None of *the cast iron is removed with the scale.*"

"The self-scaling action, which has been found to be the same in all cases where the boiler has been worked, can only be explained by conjectures, which it is not, perhaps, necessary to introduce in the present paper. It deserves the careful investigation of the chemist and mechanical philosopher, with whom the author prefers to leave the subject."

The property of casting the scale, as stated by Mr. Colburn, was not aimed at when the Harrison Boiler was designed, and such a result was entirely unexpected when it was first put in use. It is true, nevertheless, that after three months trial of my first boiler, at Messrs. William Sellers & Co., in 1859, no scale or other deposit of any kind was found therein, all that may have been formed having been removed by merely blowing out once a week. Still, this short experience seemed hardly sufficient to establish a rule.

But, in addition to what has been stated by Mr. Colburn, a continued experience, running through many successive years, in Philadelphia and elsewhere, has proved that, as a general rule, the Harrison Boiler *does* regularly shed its scale by blowing it out, under certain directions, once a week.

There are, I think, several reasons why this peculiar form of steam-generator should have this property. In the first place, cast iron, as a material for steam-boilers, has not the same tendency towards continued oxydation as wrought iron.

I do not think that the hollow sphere, in itself, has any special quality for throwing off scale; still, I think it will be seen that the hollow sphere, in connection with the curved neck, may have this quality in some degree. If the form of the interior of one of the units of the Harrison Boiler is examined, it will be found that but little of its inside surface makes a complete arch or continuous ring.

A hollow sphere, without opening of any kind, would bind deposit to its inner surface, just as it is bound in a tube of equal diameter, and from which it is often so difficult to be removed. In almost every instance in the "units" of the Harrison Boiler, the reversed curved line comes in to break up the continuous ring, and removes the abutment of the arch.

The boiler should be blown off frequently during its working hours, when broken scale and organic matter is in suspension in the water. It should certainly be blown out entirely empty under pressure once a week, after the fires are drawn and the furnace walls cooled down to a moderate temperature. On no account should a boiler be blown out while the walls are red hot. In some instances, owing to the peculiar nature of the water, a *soft* or a very *hard* deposit may occur, that will not crack off or blow out as above. To avoid trouble from this cause, it is necessary that some of the lower bolts be taken out for examination, say once in three months. This *soft* or *hard* deposit being found, the boiler must be kept clean thereafter by taking out the bolts, and clearing the deposit with a scraper adapted to the purpose, or by the use of solvents.

Interior deposit, next to corrosion, is the great evil that most seriously affects all steam boilers, and it is not pretended that the Harrison Boiler is free from this trouble.

Many schemes are resorted to for removing interior deposit, and if the same amount of time, trouble and expense, were applied in preventing the introduction of injurious matter, the evils that arise would in a great degree be remedied.

The great problem in steam, has been its generation, its use being a comparatively easy matter. Yet we find engine builders give most attention to improvement in the making and managing of the engine, leaving usually that more important thing—the boiler—to take care of itself.

Firemen or stokers, on whom rests every moment, very great responsibility, are but too often allowed to bungle along in their own ignorance and stupidity. If they can be taught to note the height of water, and to shovel coal into the furnace, this is about all that can be expected of them.

CIRCULATION OF WATER.

It is believed that the steam-boiler under consideration has many advantages in its water and steam circulation. When set, at an angle of about forty degrees, the currents of water, in each section, when generating steam, ascend along the line of spheres nearest the fire, and through the intersecting necks, carrying with these currents the accumulating steam, and delivering it at the upper angle, or steam-room portion of the section. The descending water-currents, at and near the water-level, find their way downward along the upper ranges of spheres, and through the intersecting necks, reaching at length the lower angle of the sections, from which point the circulation is continued, and goes on as before.

It should be noted that each section of this generator, no matter how large the whole structure may be, has in itself a separate and thorough system of water and steam circulation. In fact, each section is a distinct boiler.

FIRE CIRCULATION.

The boiler under consideration, it is believed, has several advantages in the application of heat thereto. Its peculiar form softens the lines of circulation for the heated products of combustion, leaving no abrupt turns or out-of-the-way corners, so often found in the more complicated boilers, and at the same time presenting a series of uniform channels, with their equally uniform surrounding surfaces, for taking up the heat. Unlike the long and narrow tube of the tubular boiler, each channel of fire circulation in this is in full connection with all the others, both vertically and laterally, causing a better blending of the elements that support combustion, and thus maintaining it for a longer period than can be done in the long narrow tube.

It has been proved by repeated experiment, that tube fire-surface (say in tubes of two inches external diameter and under) has a greatly reduced value, after the first two or three feet from the fire, when compared with fire-box surface.

In some carefully conducted experiments made under my notice several years ago, at St. Petersburg, Russia, it was found, in boiling water in the open air, that copper tubes of two inches external diameter and one-tenth of an inch thick, laid horizontally, were very nearly equal to fire-box surface, in evaporating power, for a distance of two and a half feet from the fire. Beyond that distance their value fell off rapidly, and five feet, with a most intense fire, seemed to be the maximum of really valuable fire-surface. Between the ninth and tenth foot (the latter being the extreme length of the tubes,) water could not be raised in temperature to two hundred and twelve degrees, after hours of continuous firing.

These experiments, often repeated, and always with the same result, were made with a fire-box twenty inches diameter, and twenty inches high, out of which proceeded, horizontally, four copper tubes of the dimensions above stated. Great care was taken to ascertain the exact evaporating value of each separate foot of tube-service in the direction of their lengths, as well as the value of fire-box surface.

Flame, or inflammable vapor, entering a tube of small diameter, may, at the moment of entrance, be in full combustion, and have a temperature nearly or quite equal to the source from whence it springs. Passing into a tube, one and eight-tenths inches internal diameter, surrounded by boiling water, the metal in the tube would have a temperature, at the pressure of the atmosphere, of not much above two hundred and twelve degrees. The vapor, with no addition to its supply of oxygen, coming in contact with this comparatively low temperature of the tube, would soon be reduced below the point of combustion, and from that moment, no matter what heat-giving properties it might still possess, they would be of no further effect, and to the end of the tube the vapor would give off heat only as heated air.

MAINTENANCE.

From the small size of the parts, and the ease with which they are put together or taken to pieces, it will be seen that injured parts of the Harrison Boiler can be renewed with great facility. A sphere may crack, or portions of the sections, nearest the fire, may, from lowness of water or other cause, become overheated and spoiled. The taking out of a few bolts, and the removal of a small portion of brick-work in a stationary boiler, permits the displacement of defective parts, without disturbing the uninjured portions of the structure; replacement, being equally convenient.

TRANSPORTATION.

The facility with which this boiler can be maintained, will explain why it is easy of transportation. Usually it is sent from the workshop in sections convenient and safe to handle, each weighing about a ton. These sections can be put into place, if necessary, through an opening five feet long and one foot wide.

If it is required to still further reduce the size of the separate parts, a section may be taken to pieces, so that no portion need weigh more than eighty pounds.

Thus a boiler, no matter what may be its ultimate size, can be carried in detail in a man's hand, and may, if necessary, be put through an opening one foot square. The hauling, transporting and placing of the unwieldy wrought-iron boiler is always a source of great expense,—often of danger.

Many a boiler has been run much nearer the point of disaster than it would otherwise have been, had its removal and replacement been a matter of more speedy and easy accomplishment. Many a disastrous explosion has occurred, and many a valuable life lost, for no better reason.

EXTERNAL CLEANING.

It is always difficult to keep the ordinary steam-boiler entirely free from soot and other deposit on its fire-surface, whether in flues, plain cylinders, or the other numerous forms in which it is made.

The fire-passages of the Harrison Boiler, being uniform, and all in connection with one another, external cleaning becomes a very simple process. A steam jet is best adapted to this purpose, and a convenient apparatus of this kind is attached to each boiler, so arranged that its use for a few moments every day, will keep the external surface of the spheres clean.

TENDENCY TO RUPTURE.

Steam at ordinary working pressure has little tendency to rupture any of the parts of this boiler. Neither has the steam pressure any great tendency to separate the joints, as the necks, being but three and one-quarter inches in inside diameter, might be held together, under a pressure of one hundred pounds per square inch, with a bolt no larger than would sustain a strain of about eight hundred and thirty pounds.

Allusion has been previously made to the evils that affect all boilers from irregular expansion when in use. It is equally necessary in the Harrison Boiler to guard against this influence.

It may be well here to consider this most important feature more fully.

The powerful and irresistible action of this force, begins its work at the moment fire is first applied to a steam-boiler, and little by little, too surely impairs its strength. It is most difficult to control when the material is in such form as not to admit of compensation or allowance, when subjected to its injurious effects.

In ordinary boilers made of wrought iron, it is practically impossible to arrange the parts so as to prevent irregular expansion, and consequent undue strain, from this cause. As it has been shown that it is equally impossible to make and put such boilers together when new, without undue strain, we have here two most powerful influences at work, tending to rend the parts asunder (entirely irrespective of steam-pressure), the importance of which as elements of danger, are seldom, if ever, taken into account.

Take a plain cylinder, if you please, the most simple form into which wrought iron can be put, to make a steam boiler: put a fire under such a boiler, expending its most intense heat upon a short portion of one-half of its lower diameter, at the end over the fire-grate, the products of combustion thereafter, coursing along the whole length of the remaining half diameter, until the outlet to the chimney is reached. Under these circumstances, the lower line must be very materially increased in length over the upper, and the whole structure will be then subjected to a series of complicated strains, the position and nature of which we can only conjecture, but cannot provide against. Extra thickness of material will not always remedy, and might aggravate the evil.

In more complicated boilers, this terrible ordeal is often yet further intensified. Make such boilers of brittle material,—putting them into shape, entirely free from strain (if this were possible), and let them be fired in the same manner as if made of wrought iron. That such boilers would break in pieces ere long is not doubtful. Made of wrought iron, they might not break up at once, but their greater tenacity would give them no immunity from the influences that had so soon destroyed their more brittle competitor, and which might in the end prove alike fatal to both. It is not a very hopeful view, when we consider that these undue strains are not lessened when the boiler becomes weakened by corrosion, or has its strength impaired by any other cause.

Cast iron expands less at the same temperature than wrought iron, and this difference might seem likely to interfere with the tightness of the joints in the Harrison Boiler. But with the compensating curved lines of the units, and a due proportion being maintained between the bolts and the spheres, no trouble need arise from irregular expansion.

DURABILITY.

In an experience of many years with this boiler, no serious gradual depreciation has ever shown itself in any of its parts. Corrosion affects it but slightly.

In fact, except from overheating or *"burning,"* arising from low water or other causes, and consequent warping of the "units," which, of course, destroys the accuracy of the joints, I have not found any decided marks of depreciation.

SECURITY FROM EXPLOSION.

By what has been adduced it must be seen that the Harrison Boiler is safe from destructive explosion. It is not, however, maintained that it cannot, under undue strain, be ruptured in some of its parts, or that it might not do injury, consequent upon a sudden discharge of water or steam. But it *is maintained*, that under no circumstances can it "*rend and scatter large masses of material, liberating at the same time large volumes of highly charged water and steam.*"

On page 131 of the *Journal of the Franklin Institute* for February, 1867, will be found a report of the "Committee on Science and the Arts" of the Franklin Institute, giving an account of certain severe tests that the Harrison Boiler was put to, in the effort to destroy it by steam-pressure and other means. The attempted destruction utterly failed. Attention is called to this report, at page 39, as exhibiting some very remarkable results.

When it is considered that eight hundred and seventy-five pounds per square inch of steam-pressure, failed to burst any of the spheres in one of the sections,—that under such severe test every joint becomes a safety-valve, and when it is certain that, under all circumstances, the general integrity of the whole structure can be surely maintained (a point most positively insisted upon), then but slight injury can arise, in any contingency.

CONCLUSION.

Considering the plan, material and mode of manufacture of the Harrison Boiler, let us now revert to the *axioms* laid down by me as guiding principles in making a steam-boiler, and see how far they have been carried out.

1st. The boiler under consideration *is theoretically and practically safe from all destructive explosion, even when carelessly used.*

2d. *It is constructed upon a system or series of uniform parts, few in number, easily made, and easily put together or taken apart, and not of costly material.*

3d. *Its strength is in no respect dependent upon any system of stays or braces, whereby the inefficiency or rupture of one of these braces or stays can cause greatly increased strain upon the others, thereby endangering the whole structure.*

4th. *Its parts are not of great weight or size, and may be easily transported, or put into or be taken out of place.*

5th. *It has a principle of renewal, allowing the easy displacement and replacement or interchange of any and all of its parts, without impairing or disturbing the material or workmanship of portions not needing renewal.*

6th. *Whether of large or small size, it has uniformly such elements of strength, as will always render it capable of sustaining many times greater pressure than need ever be demanded of it in practice, and its safety is not impaired by corrosion, or the many other harmful influences that so soon and so seriously affect the strength of ordinary boilers.*

7th. *Its parts are so made and put together that in case of rupture, no general break-up can occur. Its contents may be discharged, but no explosion or serious disturbance of any kind can take place, consequent upon such discharge.*

8th. *It is constructed so as to facilitate the removal of deposit of all kinds, from its interior and exterior surface.*

It is assumed that the better "guiding principles" in the construction of a steam-boiler have been fairly carried out,—that all other matters are at least equal, and that the result *is* a safer, if not a better one than any that has preceded it.

ITS PRACTICAL AND COMMERCIAL SUCCESS.

Until the spring of 1864, no effort was made to test the commercial merits of the Harrison Boiler by offering it for sale. In the early part of 1863 a boiler of fifty horse-power was put up experimentally at the establishment of Messrs. John Hetherington & Sons, Manchester, England, and subsequently a smaller one, at the same place. Both of these boilers worked well and gave great satisfaction up to the middle of 1864. Orders were also received for several others, to be erected in the neighborhood of Manchester. Several hundred tons of this boiler, made in England, were imported into this country in 1864, and put into operation in Philadelphia and other places.

Since October, 1865, I have been manufacturing the boiler at my own foundry on Gray's Ferry Road, Philadelphia. There are now at work, in the United States, boilers on this plan, varying in size from five to four hundred horse-power. Many of these have been erected in Philadelphia. Others in various parts of the Union, from Maine to Texas, and from Massachusetts to San Francisco. Some have gone to Canada, others to South America, Liberia and New Zealand.

The advantages of the "Harrison Boiler" were understood and appreciated at the International Exhibition held in London in 1862, at which time a first-class medal was awarded to this invention, in Class VIII, *"for originality of design and general merit."*

The Harrison Boiler has met with more favor at the hands of the public than could have been expected, considering the novelty of its material and form, and the prejudice that so naturally attaches to any effort aiming at an almost entire overthrow of a long established system. Its peculiarities invite criticism, and it would not appear strange if even many of those best acquainted with the subject generally, should pass this by with little attention, thinking at a glance that it was so much out of the beaten track, as to seem utterly impracticable.

Nothing in connection with the use of steam has been so much discussed, as the manner of making the apparatus for its generation, and to have called out, for a century past, such an army of thinkers on a subject having in the abstract but a few simple elements, there must have been, and no doubt still is, some inherent defect in the plans heretofore and now used.

If, in these pages, I have added to the stock of information tending to make the use of steam less dangerous to life and property, I have attained something. If I have been instrumental in producing a steam-boiler, that will take its place permanently, as a means of rendering steam-generating apparatus more safe from destructive explosion, I shall have attained something more. If in my effort to improve a much used and much abused object, manifestly demanding improvement, I have only succeeded in proving a fallacy, I shall still have my reward.

JOSEPH HARRISON, Jr.,

Rittenhouse Square, Philadelphia.

DIRECTIONS FOR USING

THE HARRISON STEAM BOILER.

IN screwing up the Harrison Boiler, the power of one man on a three-foot lever is all that should be put upon the bolts. No amount of screwing will avail if the parts have been warped by overheating.

As with all boilers having small water capacity, compared with evaporating power, this requires careful attention to the water-level. It is placed in this respect with the locomotive tubular boiler, and requires the same treatment. Too low water is equally harmful to the Harrison Boiler as to all others.

It is sometimes urged that the Harrison Boiler has too little steam-room. It rarely happens that this objection is valid. In cases where large volumes of steam are required, at irregular intervals, it may be necessary to keep a stock on hand; but this can only be done at a loss of pressure or a loss of heat. If extra steam-room is required, it can be had to any extent by adding to the upper portions of the units. If plate-iron drums are resorted to, they may be arranged without difficulty.

The Boiler should be blown out under full pressure, until empty, at least once a week, after the fire is drawn and the walls cooled below red heat. It should not be refilled with water until quite cold. By following this direction, it will, under ordinary circumstances, be kept free from injurious deposit or scale.

When the Boiler is divided into two or more sections, or worked in connection with other boilers, then blow steam and water entirely out of one section under full pressure, retaining steam under full pressure in the other; and before closing blow-off cocks, turn full head of steam into the empty boiler. This rush of steam through all parts of the discharged boiler, if continued for some time, will often drive out all loose scale or soft matter, and do more to prevent injurious deposit than any other means.

In some instances, owing to the peculiar nature of the water, a soft or hard deposit will accumulate, and crack off, but not blow out as above. In this case, it will be necessary to remove the lowest bolt in each slab for examination at least once in three months, and to clean out the deposit with a scraper.

Where water impregnated with lime is used in the Harrison Boiler, it is advisable to use something to loosen the scale, and to prevent it from forming. We have tried many of the patented anti-incrustation powders and liquids in vogue for this purpose, but have not found anything equal to or better than *Crude Petroleum.*

It may be poured in at the safety valve when steam is down, or forced in with the water by the pump. The quantity of oil to be used depends upon the character of the water.

We can supply a suitable receiver to attach to the feed pipe, so that it can be forced into the boiler, either by the pump or by an injector.

It is not fair usage with the Harrison Boiler, or any other, to run it night and day, without intermission, for a week. Scale is thrown off from the inner surfaces by slight differences of temperature between the material of the boiler and the inside deposit of earthy or chemical matter, the latter being slightly broken by expanding or contracting less than the metal upon which it rests. To enable this difference of temperature to take place, it seems necessary that the boiler should be at rest, or nearly so, for a portion of the twenty-four hours of each day. Boilers running for six consecutive days of twenty-four hours each, and only cooled on the seventh day, may collect so thick a deposit that it will not scale or crack off with a single change of temperature. Under these circumstances any boiler would require the deposit removed by mechanical or other means, before it becomes so thick as to cause the material of the boiler to be unduly heated.

For the greatest efficiency in generating steam and economy in fuel, it is necessary that the external surface of the spheres should be kept at all times as clean as possible from ashes and soot. The best way to do this is by a steam-jet—an apparatus supplied, when wanted, with the

Harrison Boiler. By using this jet for a few minutes, through the small doors in the fire front, once in twenty-four hours, exterior deposit can be easily removed.

When double sets of boilers are used of any kind—a single set being amply necessary for doing the work—then, unless in cases of emergency, such as repairs or cleaning, experience has shown, that economy in fuel, and durability in boiler and fittings, run entirely in the direction of working both sets of boilers all the time, or as near so as may be. Low fires induce low temperature in the chimney, a sure test of economy in fuel; and although the area of grate may be doubled, the saving will be surely found in the bills for coal. Working a boiler too hard, like overworking a good horse, tends to disablement. Boilers not used have a tendency to depreciate, just as horses are not benefitted by standing idle in the stable.

HARRISON BOILER WORKS,
Gray's Ferry Road, adjoining the U. S. Arsenal,
Philadelphia, Pa.

February, 1871.

REPORT*

OF THE

COMMITTEE ON SCIENCE AND THE ARTS,

CONSTITUTED BY THE FRANKLIN INSTITUTE,

ON THE

HARRISON STEAM-BOILER.

Invented by Joseph Harrison, Jr., Philadelphia, Pa.

THE Committee to whom was referred the examination of the "Harrison Boiler," report that, on Tuesday, October 30th, 1866, they visited the foundry of Mr. Joseph Harrison, Jr., Philadelphia, and had an opportunity of inspecting the boilers in various stages of manufacture, and of seeing several in operation.

Experiments were tried to prove the strength and durability of the boiler, under extraordinarily severe use.

These boilers are of cast iron, formed of a combination of hollow spheres, each eight inches diameter externally, and three-eighths of an inch thick, connected by curved necks three and one-quarter inches diameter. These spheres are held together by wrought iron bolts, and in one direction are cast in sets of two, or four, with opposite lateral openings to each sphere, and are called by the inventor two or four-ball units, as the case may be.

He assumes that the boiler, in its smallest form, may be considered as one of these balls, with its opposite lateral openings closed by caps held in place by bolts. Two balls united by a neck, with caps over the four lateral openings, in the same manner would also make a boiler of a larger size. Four balls so united in one casting, would be a still larger boiler, and that any number of these balls or spheres may be united by bolts passing through them so as to form large boilers, and

* See Journal of the Franklin Institute for February, 1867.

the strength of the boilers so made, will be the strength of the weakest sphere or ball in the structure.

In manufacturing the boiler for ordinary use, a number of these units are generally so arranged, as to form sections twelve and thirteen balls long, six balls wide, as shown by the annexed sketch A. These sections are all tested by hydrostatic pressure, as high as three hundred pounds per square inch, before being delivered to purchasers. The Committee saw one of these sections subjected to a bursting pressure of water, one sphere bursting when the pressure had reached six hundred pounds per square inch. A second one tested in the same manner, burst at six hundred and twenty-five pounds. They were shown a section in which one unit had burst at nine hundred pounds per square inch, the damage

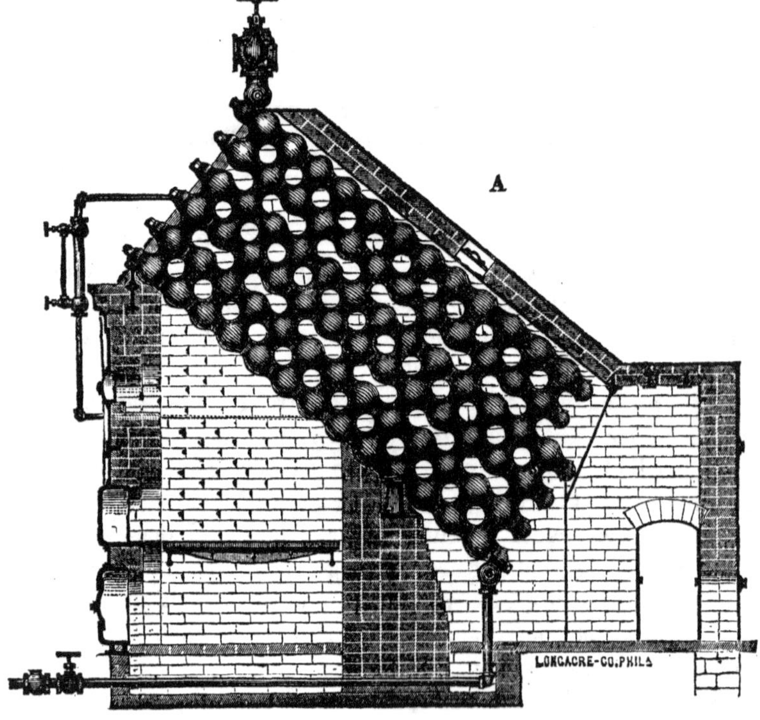

having been repaired by the insertion of a new unit. The section then stood eleven hundred pounds per square inch before bursting in a new place. The available strength of the section in all cases being the strength of the weakest unit in it, the inventor holds that the boiler is safer than any other in use; in fact, he considers it entirely free from any danger of disastrous explosion. To prove which, he had a section equal to six horse-power, similar to the one tested by hydrostatic pressure, and such as he is regularly selling, placed in an extemporary furnace built in a clay bank, and set in the usual manner for a boiler of this kind.

The boiler was filled with water to the regular height, say about two-thirds full, with no outlet or safety-valve of any kind, and sealed up tight, a small tube leading from the upper ball to a high pressure-gauge, placed at a safe distance, about two hundred feet, from the boiler. A fire was made under and around the boiler, with the fuel of dry pine wood. The wind was very high at the time of the experiment, blowing from the west directly into the furnace, thus fanning the flames to an intense heat.

The gauge soon gave indication of the formation of steam, the pressure steadily increasing up to four hundred and fifty pounds to the square inch.

At this pressure there seemed to be a sudden discharge of steam, as from a small opening. The discharge did not continue for many seconds, and the Committee are not certain that it proceeded from the boiler; there may have been some water discharged from the bank of wet earth into the fire. The pressure then increased at an uniform rate until it had reached the enormous strain of *eight hundred and seventy-five pounds per square inch*, when a sudden discharge of steam took place, seemingly no greater in volume than might issue from a safety-valve of two and a half inches diameter, or even less; after which the pressure fell to four hundred and fifty pounds, at which it stood when the fire was drawn for examination. While this boiler was being uncovered for examination, a boiler of about twelve horse-power, consisting of two sections, similar to the one previously experimented upon, was fired, and steam raised to one hundred and twenty-five pounds pressure. This boiler had no safety-valve, but was provided with a globe-valve of one inch capacity or area, as an escape-valve to regulate the pressure in the boiler. When the Committee examined this boiler at time of firing, it had two full gauges of water. The escape-valve was opened so as to reduce the pressure to one hundred pounds per square inch, and regulated from time to time to keep the pressure uniform at this point. The fire was pushed, and no more water injected into the boiler. In due time the lowest gauge-cock gave no indication of water. Soon afterwards a slight leak was observed in one joint of the left-hand section. This closed in a few minutes, and one opened in a similar manner in the right-hand section; this also closed in a short time. No other leaks showed themselves during the experiment. As the water boiled away, the soot began to burn off the upper balls of the sections, that is, off those of the upper balls of the lowest row, visible through a peep-door above the fire-door provided for inspection. The boiler then gradually became red-hot, and even when all the water seemed to be exhausted, and the pressure slowly fell, the gauge stood for some minutes at thirty pounds, as if from the vaporization of some water in the lower courses of the sections, showing that in this red-hot condition, the boiler was tight enough to hold pressure. After the fire had been drawn, and the boiler cooled, the bolts holding the units together were found to be loose, as if stretched by the unusual heating of the cast iron surrounding them. During the time of the experiment with low water, the escape-cock was many times closed to increase the pressure, then opened quickly to reduce it to the one hundred pound standard, but with no deleterious result. When the gauge stood at thirty pounds, all of the boiler visible from the peep-door and fire-doors, down to the bridge-wall of the furnace, was at a bright red heat. This was unmistakable, as, when the fire was drawn, the boiler was hot enough to ignite wood held against it.

November 13th, 1866.—At four o'clock, P. M., the Committee met at the factory. J. Agnew and J. C. Cresson present. They examined the boilers tested at the former meeting. The boiler which had been subjected to its own steam-pressure of eight hundred and seventy-five pounds per square inch, had been removed to the factory for examination. Mr. Harrison's foreman stated, that when the boiler was first dragged from the fire, after its water had been forced out, (as detailed in the account of the experiment,) the three lower bolts were quite slack, but the next morning, when it had become cold, one of them was again tight. The other two were not quite tight, but were then screwed up about one turn of the nuts. The Committee are confirmed in their belief that in this extreme test, the pressure at eight hundred and seventy-five pounds, was enough to stretch some of the bolts, that the joints opened as safety-valves, and thus relieved the strain on the boiler.

The boiler which, in former experiments, had had all its water boiled out, and had then been heated to bright redness, was found to be quite sound and fit for use, making steam freely, and showing no leak, blowing off at sixty-five pounds by the safety-valve. It was somewhat disfigured on its outside, by oxidation. Your Committee was informed

NOTE.—Notwithstanding the immunity in all respects with which these boilers withstood the unusual and most severe tests related above, still it is imperative that the proper height of water should be always maintained in the Harrison Boiler, as too low water is as likely to do injury to it as to any other.

that it had not been changed or repaired since the trial, but that some of the bolts had been screwed up.

A third boiler of the same size as the above, twelve horse-power, was then tested in the following manner: After being filled with water to the upper water-line, it was fired until pressure was raised to ninety pounds, at which it was blowing-off freely. The water was then all blown out by the blow-off cock, the pressure falling to sixty pounds while blowing-off, at which it stood until steam reached the blow-off pipe, when the pressure fell to zero. It was kept empty for three minutes, with the fire still burning, and was then rapidly filled with cold water, and steam raised to one hundred pounds pressure in thirty minutes, blowing off at one hundred pounds, and was quite sound and tight.

The Committee was informed by one of its members, who was a witness of, and cognizant of, all the facts, that at the establishment of Mr. Wm. Sellers & Co., of this city, a boiler of this kind has been in use for about two years. During some experiments in testing the Giffard Injectors made by that firm, a workman inadvertently loosened a connection to the water supply-pipe, resulting in the pipe blowing full open, discharging the water from the boiler as fast as a two-inch diameter opening would allow, the men in the boiler-room barely escaping with their lives. As soon as all the water had blown off, and access could be had to the boilers, the fires were drawn and cold water run in as fast as possible, and in about thirty minutes, the steam was high enough to run the engine, with no seeming injury to the boilers.

The Committee mention this as an accidental experiment, similar to the one above reported. The same boiler is still in use, and seemingly as good as when first erected. It is, however, the first one erected in this country from units made in England, and is not so good as those made since then. On Saturday, November 17th, Mr. Harrison repeated an experiment in the presence of a part of the Committee, Messrs. Agnew, Morton and Sellers, which experiment he stated had been tried twice the day before, and once two days previous, all the experiments being with the same boiler. The experiment, as witnessed, was as follows:

The boiler which had been under experiment November 13th was fired up, and steam raised to one hundred and ten pounds. The fire was active,—what might be called a very clear fire,—and in good condition to make steam freely. It had been kept up sufficiently long to thoroughly heat all the furnace walls. Steam was blowing-off freely from the safety-valve. At a given signal the blow-off cock was opened suddenly, blowing off all the water until the pressure had fallen to zero, and neither steam nor water was escaping from the blow-off cock. In fact, it is believed the boiler was entirely dry. The blow-off cock was then closed, and cold water from a well pumped rapidly into the hot boiler, for it was at all times exposed to the active fire. As the water entered the boiler, the pressure as per gauge, rose slowly during an interval of about three minutes, when it is supposed the water-level had reached the more heated portion of the boiler above the bridge-wall of the furnace, for the pressure seemed instantly to increase to one hundred and ten pounds, and steam blew freely from the safety-valve.

This pressure and escape of steam, continued for some minutes with no variation, when suddenly an escape of steam was evident from the boiler into the furnace, and upon opening the peep-hole door a jet of water was seen issuing from one of the joints. This leak, in less than a minute, suddenly stopped; then, as the water rose in the boiler, a similar sudden leak and sudden stoppage, occurred at the next higher joint; again, at a third one, when, by that time, the water was showing itself at the lower gauge-cock, soon afterwards at the second one, when the pump was stopped, at which time the pressure stood at one hundred and ten pounds, steam blowing off freely from the safety-valve. The fire was as active as when the experiment began, and the boiler perfectly tight. This experiment, as before remarked, had been repeated three times previous to the one witnessed by the Committee, and Mr. Harrison's account of the previous experiments, given to your Committee, agreed in every respect with the facts as seen by them. This is as severe a test as any boiler is ever accidentally caused to sustain, and is, in fact, the one most likely to occur from carelessness. It is also testing practically, the favorite theory to account for explosions. During the experiments, the employees of Mr. Harrison seemed quite fearless in their manipulation of the boilers, showing a confidence in their safety, truly remarkable. With the exception of the single boiler sealed up and submitted to the extreme pressure of eight hundred and seventy-five pounds to the square inch, all the experiments were tried within the building in which the boilers are made, and any explosion would have resulted in serious loss of property, if not of life. Had any ordinary wrought iron boiler, made in the simplest form, and of the best material, been submitted to these same tests, it would have probably been destroyed by any one of them. Regarding the liability to accumulation of sedimentary deposit in this kind of boiler, we can only say that it is asserted by those who have used them the longest, that by occasionally blowing out the water under a full head of steam, then allowing the empty boiler to be moderately heated by the hot furnace, filling up with water and rinsing out, the scale becomes detached and rushes out at the blow-off cock.

The Committee have carefully inspected the manner of making these boilers as practiced by Mr. Harrison, and find the greatest care is taken to insure perfection of workmanship; but, at the same time, it is eminently noteworthy, that the peculiarities of the boiler, and its mode of manufacture, are such as to enable a high degree of mechanical excellence to be obtained by mechanical devices, apart from the workman's skill. Thus, in the process of casting, taking as an example a four-ball unit, the four eight-inch spheres united by necks three and a quarter inches diameter, internally, have on each ball two opposite lateral openings, three and a quarter inches diameter, thus making in all, eight openings to four balls. The patterns are all of cast iron, parted lengthwise through the centre of the unit, by a plane at right angles to the lateral openings, these serving as supports to the green sand core which is moulded within the pattern itself, and not in a separate core-box, thus insuring absolute uniformity to the thickness of the metal, and offering a more yielding core to the contracting metal than in the case of dry-sand moulding. The lateral necks which are to serve as joints in combining the units into the boiler structure, are faced off by machinery of the most ingenious kind, so arranged as to insure neat accuracy in the surface, the joints on one side having depressions to match projecting tongues on the other, these tongues serving with the longitudinal bolts to hold the units in position. One of the most thorough descriptions of this kind of boiler is the report of a paper read by Mr. Zera Colburn before the Institute of Mechanical Engineers in 1864, an abstract of which can be found in *Engineering Facts and Figures*, by A. Betts Brown, for 1864. He shows, that although the tensile strength of cast iron is not so great as wrought iron, yet the spherical form of each unit of the boiler gives it an equivalent strength. He says: "The strength of a hollow sphere to resist internal pressure, is exactly twice that of a hollow cylinder of the same diameter, material and thickness, and it can be shown that even a cast iron sphere, seven feet in diameter and seven-sixteenths of an inch thick, is as strong as the shell of a Cornish boiler of the same dimensions." "The plane in which rupture, if it happen at all, will take place in a hollow sphere, is the largest plane that can be drawn through it, and the metal resisting the strain tending to cause rupture, is the whole section of metal bounding the plane." "In a hollow cylinder, the area upon which the greatest pressure tending to cause rupture, will be exerted, is that represented by the product of the length into the 'diameter of the cylinder.'" The ends of such a cylinder, add nothing to the strength of the cylindrical part, in case of a rupture beginning at the cylindrical part.* The spherical form of each part of this boiler is one of its marked advantages, not only so far as strength is concerned, but as enabling a much larger amount of surface to be exposed to the fire than in any form of combined cylinder. To the spherical form with the curved necks, has been ascribed by the inventor the property, which this boiler is asserted as having to cast its scale when emptied of water, as there is no seeming abutment for the arch of the crystallized scale to spring from. The value of cast iron, so far as durability is concerned, has long been conceded. The purer the iron the more readily does it corrode, while the mixture of even a small amount of carbon increases its ability to resist corrosion. Wrought iron water-pipe under ground soon rusts out. Cast iron, even of the same thickness, remains good after many years' use; in fact, is considered practically to suffer no deterioration. Wrought iron in boilers decays internally—the most rapidly where moisture and air both operate, as in the upper side of mud-drums, while they are often eaten through from the outside by trifling leaks, and the constant trickle of water over the surface. Wrought iron boilers are, according to the experiments of Fairbairn and others, so much weakened by the process of riveting, &c,. as to suffer a deterioration of about forty per cent. The Harrison boiler is made of pieces of as uniform strength as possible, united in a systematic manner. The uniting the units or pieces into mass, does not diminish their strength. In case of accident to any part of the boiler, the damaged part may be removed, and instead of being repaired, as is done with wrought iron boilers, new parts may be substituted, just as bricks may be taken out, and new ones replaced in a building. The patching of a damaged wrought iron boiler makes it weaker. The renewal of any part of the Harrison boiler gives it its original strength.

The experiments heretofore described, have been conducted to de-

* The metal effectively resisting the rupture in the cylinder, being only the length of the cylinder. Thus, by comparison, Mr. Colburn arrives at his conclusion as to the relative strength of the two forms. (See *Engineering Facts and Figures*, 1864, pages 12 and 13.)

termine the safety and durability of the boiler under unusual and severe usage, or rather to determine whether any danger can result from submitting this kind of boiler to those circumstances which, in ordinary wrought iron boilers, are thought to result in explosions, or great injury to the boiler.

The Committee are impressed with the great utility of the boiler, as one *perfectly safe and free from all danger of explosion, even when carelessly used.* This recommendation alone, in a humanitarian point of view, must strongly commend it to public favor. During the experiments, its steam-making qualities were favorably noticed, and such boilers in actual use as your Committee have had an opportunity to examine, seem to give satisfaction in point of economy; but in the absence of all experiments in this direction, conducted under their immediate supervision, they do not feel qualified to report in figures as to its steam-making efficiency.

Comparing cast iron plates with wrought iron ones of the same thickness, the transmission of heat is known to be in favor of the former; hence the material, if in a safe form, is better adapted to economical steam-making, than wrought iron. Ordinary boiler plate is seldom less than one-fourth of an inch thick, and more commonly three-eighths, particularly for high pressure. The castings used in the experiments for safety, were not over three-eighths of an inch thick, and in one boiler set up in a form adapting it to marine purposes, some of the units were only three-sixteenths of an inch thick, and were worked successfully at one hundred pounds pressure, driving all the machinery in Mr. Harrison's factory in an efficient manner. The principle of enlargement of the boiler by addition of units, and the fact that it can be constructed in any shape or style, just as various kinds of buildings are constructed of ordinary bricks, places it in the power of the engineer to adapt it in its form to the requirements of each particular case; so that with the known advantage of the use of cast iron, and the unlimited scope in the arrangement of heat-absorbing surface, coupled with the demonstrated fact of safety, your Committee unhesitatingly *approve, and heartily recommend it to public favor.*

Sub-committee appointed to make the examination: Coleman Sellers, Chairman; John Agnew, John F. Frazer, Henry Morton, J. C. Cresson.

NOTE.—Notwithstanding the immunity in all respects with which these boilers withstood the unusual and most severe tests related above, still it is imperative that the proper height of water should be always maintained in the Harrison Boiler, as too low water is as likely to do injury to it as to any other.

PARTIAL LIST

OF

SALES OF HARRISON BOILER.

PENNSYLVANIA.

Philadelphia—At Wm. Sellers & Co.'s Foundry; S. W. Cattell's Woolen Mill, 2 boilers: Stephen Robbins' Rolling Mill; Taws and Hartman's Brass Foundry; Girard Flouring Mill, Ninth Street; G. W. Simons, jeweller, Sansom street; James B. Rodgers Co., Printers; H. C. Oram and Co.'s Foundry; Henry Bower's Chemical Works, 2 boilers; Duplaine, Cooper & Stokes' Rolling Mill, Richmond; Geo. Lutz; Dr. Joseph H. Schenck, Pulmonic Syrup; James Yocum & Son's City Iron Foundry; The *Evening Bulletin* (Printing Press); Winsel & Pearce, Manufacturing Jewellers; Con. Cline, Vine Street; H. C. Fox, Glass Works; Robert Shoemaker & Co.'s Paint Mills; L. Martin & Co.'s Paint Mills; L. Martin & Co.'s Chemical Works; James, Kent, Santee and Co.'s Dry Goods House; John G. White & Son's Malt House; Powers & Weightman's Chemical Works; Charles Lennig's Tacony Chemical Works; Henry Disston's Saw Factory, 2 boilers; E. Ketterlinus' Printing House; Hospital of the Protestant Episcopal Church; David France's; Tatham Brother's Lead Pipe Works; Wilbank & Howard's Machine Shop; Thornton and Maybin's Woolen Mill; Thos. E. Lutner's Blacksmith Shop; J. Reigel's Dry Goods House; Hance, Griffith & Co.'s Chemical Works; Smith & Harris' Phœnix Plaster Mills, 3 boilers; Baeder & Adamson Glue Works, 3 boilers; William McArthur's Kindling Wood Mill; Peter Schemm's Brewery; Girard College, 2 boilers; Sigmund Rutschmann's Machine Shop: Reeves & Eastburn, Cabinet Ware Works; David Allen's Dye House; Forrest Building; Benjamin H. Shoemaker's Glass Warehouse; Edwin A. Thomas' Soap Manufactory; John F. Zehender; James Dougherty's residence; Daniel Allen, Dyer; Davis & Elverson, publishers of *Saturday Night;* Samuel S. White's new building, Twelfth and Chestnut streets, 3 boilers; Keystone Public School; James S. Murphy, 1210 Shippen Street, Dyer; Coffin & Altemus, 220 Chestnut Street; Thomas H. Powers, 1224 Chestnut Street; Frederick Lack, 510 Otis Street, corner of Holman; Mercantile Library, Tenth st., above Chestnut; Corn Exchange Building, Second Street above Walnut; Charles T. Hallowell, 1026 Ridge Avenue; Philip Gucke's, 824 St. John Street; Mitchell & Tate's, Twelfth and Callowhill Streets.

West Philadelphia. The Pennsylvania Hospital for the Insane; Murphy & Allison's Car Factory; Henry Gehring's Brewery; Gross and Brothers; Woolen Mills of Samuel Yewdell's Estate, 2 boilers; Philadelphia City Passenger R. W. Co.'s shops; Andrew Scheible's; Conrad Swartz's; Market Street Passenger R. W. Co.'s Depot; Hestonville Passenger R. W. Co.; Mr. Kreitzer, Hestonville; Herbert and Milne Planing Mills.

Altoona. Pennsylvania R. R. Co., Logan House, 2 boilers.

Germantown. (Philadelphia). Selsor, Cook & Co.'s Tool Factory; Elias Birchall, Armat Mills; F. W. & T. S. Henson, Woolen Manufacturers.

Manayunk (Philadelphia). Manayunk Pulp Works, 4 boilers; Bolton Winpenny's Mills; Stafford and Co.'s Woolen Mill.

Media. Delaware County Prison.

Lehigh Gap. Mauch Chunk Slate Co.'s Works.

Kittatiny Station, L. V. R. R. C. Foster.

Downington, Chester County. John McClure, Jr., and Brother's Woolen Mill.

Slatington. Continental Slate Co.

Reading. R. R. Co.'s Rolling Mill.

Doylestown, Bucks County. Rufe and Scheetz's Tannery; W. W. H. Davis; Doylestown Water Works; Charles Finney.

Lebanon. Lebanon Furnaces.

Adamstown, Lancaster County. Henry Stauffer's and Co.'s Hat Manufactory.

Harrisburg. State Lunatic Hospital.

Newport. Newport Manufacturing and Building Co.'s Mills.

Birmingham, Huntingdon County. Keystone Zinc Co.'s Works.

Pittsburg. Park Brothers & Co.'s Black Diamond Steel Works; Pennsylvania R. R. Union Depot Hotel, 2 boilers.

Monongahela City. E. T. Cooper.

Hecktown, Northampton County. Simon D. Stuben's Grist Mill.

Johnstown. Cambria Iron Works.

Bridgeport. Chas. Whitman.

MAINE.

Lewistown. D. F. Noyles' Card and Belt Manufactory.

Portland. Leathe & Gore's Soap Factory; Portland Stoneware Co.'s Works; W. K. Lewis & Bro.; Daniel Winslow & Sons, 4 Boilers, Portland Gas Light Co.'s Works.

Yarmouth. A. C. Young.

Pittsfield. Going Hathorn.

Lisbon. Farnsworth & Co.

Hartland. A. Linn.

Saco. York Mills, 6 boilers.

Farmington. Wm. H. Dyer.

NEW HAMPSHIRE.

Concord. Abbott, Downing & Co.

Nashua. Nashua Lock Co.'s Works; Eaton and Ayer; Pine Valley Co.'s Works.

Newport. Coffin & Nourse.

Newmarket. Newmarket Manufacturing Co., 4 boilers.

VERMONT.

Hydeville. Eagle Slate Co.
Woodstock. B. F. Standish.

MASSACHUSETTS.

Lowell. Lowell Bleachery, 4 boilers.

Boston. Standard Sugar Refinery, 750 H. P.; Dunbar, Waters & Co., *Daily Advertiser* Building; Spencer Rifle Co.; Sears' Estate, new building, Washington Street; American House; G. W. Wales & Co.; W. K. Lewis & Bro.; New England Chemical Works; Beals, Green & Co., publishers, *Boston Post;* Whiton, Brother & Co.; Farringtons, Tozier & Hall.

South Boston. Continental Sugar Refinery, 2 boilers.

Chelsea. C. F. Austin & Co.'s Bakery.

Charlestown. Union Sugar Refinery, 6 boilers.

Lynn. Holbrook & Glazier's Machine Shop.

Neponset. S. S. Putnam's & Co.'s Nail Works.

Newburyport. Merrimac Arms and Manufacturing Co.'s Works.

Newton Lower Falls. Lemuel Crehore & Co.

Northbridge Centre. Joel Batchelor's Machine Shop.

Plymouth. Plymouth Cordage Co.'s Works, 6 boilers; Standish Cotton Mills.

Rehoboth. The National Brick Co.

Taunton. W. R. Potter; Whittenton Mills, 6 boilers; Taunton Foundry and Machine Works; H. Field & Son.

Westbrookfield. George Robinson & Co.

Worcester. T. K. Earl & Co.'s Mills; Sargent Card Co.'s Works; A. N. Barrett, Washburn & Co., Gas Fitters; Thomas H. Dodge.

Walpole. Willard Lewis' Carpet Wadding Works.

Malden. Boston Rubber Shoe Co., 6 boilers.

Woburn. Winn, Eaton & Co., 2 boilers.

Milford. Clement and Coleburn.

Graniteville. Abbot Worsted Co.

Manchaug. Manchaug Co.

Winchester. A. Mosely.

Lawrence. Russell Paper Co.

Fall River. Wetamoe Mills, 8 boilers.

CONNECTICUT.

Hanover. Ethan Allen.

South Manchester. Cheney Bros., Silk Mills, 7 boilers.

New Britain. Judd & Blackslee's Hardware Works; Humason & Beckley Manufacturing Co.

Baltic. James Petrie, Anchor Mills.

Ansonia. Messrs. Wallace & Son; Ansonia White Lead Works.

New London. New London Horse Nail Co.

Jewett City. J. & W. Slater.
Norwich. A. H. Hubbard & Co.
New Haven. Moses Seward.
Taftville. Orray Taft Manufacturing Co.
Naugatuck. Connecticut Cutlery Co.
Walcottville. Excelsior Needle Co.
Waterbury. Benedict and Burnham Manufacturing Co.

CALIFORNIA.

San Francisco. Tay, Brooks & Backus.

RHODE ISLAND.

Hope Valley. Nicholas & Langworthy's Foundry.
Johnston. Thomas H. Hughes.
Providence. Gorham Manufacturing Co., 2 boilers; T. P. Shepard & Co.'s Chemical Works, 2 boilers; A. Burgess & Son's Leather Findings Mills; Silver Spring Bleachery, 8 boilers; L. J. Wood & Co.'s Chemical Works; Wm. Fletcher; Dexter Asylum, 2 boilers.
River Point. Lippit Manufacturing Co.
Bristol. National Rubber Co.

NEW YORK.

New York City. A. T. Stewart's new building, 10th and Broadway, 8 boilers; Tatham & Brothers' Lead Pipe Works, 82 Beekman street; *New York Times* Building; Mandeville & Sigler's Mill, 241 & 243 E. 23d street; *Evening Telegraph* Building; Dederick, Sears & Co., 18 Maiden Lane; Robert Bonner, publisher, *New York Ledger;* Messrs. Tiffany & Co., 15th and Broadway.
Albany. The Convent of the Sacred Heart; Albany Gas Light Company.
Binghampton. New York State Inebriate Asylum, 2 boilers.
Brooklyn. W. E. Doubleday & Co.'s Straw Goods Works.
Schagticoke, Rensellaer County. Schagticoke Linen Mills.
Sing Sing. United States R. R. Screw Pike Co., 2 boilers.
Williamsburg. Charles Illig's Brewery.
Port Richmond. William Lissender.
Poughkeepsie. D. Scott & Son.
Hoboken. Stevens' Institute.

NEW JERSEY.

Bordentown. McPherson, Willard & Co.'s Union Steam Forge.
Camden. The Camden & Atlantic Railroad's Machine Shops; Potts and Klett Chemical Works, 3 boilers; Garrison, Gillingham & Co.'s Saw Mill; Fullarton and Hollinghead's Machine Shop; Jesse W. Starr & Son's, Foundry.
Dover. The Irondale Mines.
Ellwood. McNeil, Irving and Rich's Paper Manufactory.

Elizabeth. Crane, Tubbs & Co.'s Sash Mill; Richardson & Emory; Elizabethtown Gas Light Company's Work.

Florence. R. D. Wood & Co.'s Works, 2 boilers.

Gloucester. Samuel Raby's Gingham Mills, 2 boilers.

Harrisville. Harris & Newhall's Paper Manufactory, 2 boilers.

Jersey City. United States Watch Company's Works, 2 boilers; Matthiessen & Wiecher's New Jersey Sugar Refinery, 5 boilers.

Lambertville. Lambertville Manufacturing Co., 4 boilers.

Millville. Wood & Garrett's Cotton Mill.

Mount Holly. Mount Holly Water Co.; T. H. Risdon's Machine Shop.

Newark. Walker & Sneden's Works; Ward, Huntingdon & Co.'s Saw and Planing Mills; Wright & Smith, Machinists.

Orange. T. S. Root.

Rockaway. Glendon Iron Co.'s Works, 6 boilers.

Westfield. Sandford & Moffitt's Sash and Blind Mill.

Woodbridge. Crossman, Clay & Manufacturing Co.'s Works.

New Durham. C. F. W. Meyer's Son.

Morristown. Greenwood & Hays.

May's Landing. R. D. Green.

Bridgeton. D. B. & W. C. Whitekar's Foundry.

DELAWARE.

Wilmington. Hartman and Fehrenbach's Brewery.

MARYLAND.

Havre de Grace. Craig & Blanchard's Saw Mill, 2 boilers.

Royal Oaks. J. A. Robinson's Saw Mill.

Baltimore. Welby, French & Co.

VIRGINIA.

Alexandria. J. G. Verplanck and Co.'s Saw Mill.

WEST VIRGINIA.

Martinsburg. Frazier and Clippinger's Sash and Door Mill.

Weston. West Virginia Hospital for the Insane.

Greenbriar White Sulphur Springs. Messrs. Peyton & Co.

DISTRICT OF COLUMBIA.

Washington. Government Hospital for the Insane.

NORTH CAROLINA.

Charlotte. Rock Island Manufacturing Co.'s Woolen Mills.

GEORGIA.

Savannah. Cunningham & Purse.

Penfield. Penfield Steam Mill.

Athens. Athens Foundry and Machine Works; Pioneer Paper Mills; Georgia Factory; Princetown Manufacturing Co.

OHIO.

Cincinnati. Cincinnati Type Foundry; Greenwood Pipe Co.; A. Bepler, 91 E. Front Street.

Hillsboro. C. S. Bell's Foundry.

Springfield. Charles Rabbit's Woolen Mill.

Toledo. Charles V. Brinkeroff.

INDIANA.

La Porte. Moore, Dale & Co.; Purdy, Barnes & Weir, Furniture Manufacturers, &c.

Jeffersonville. Charles C. Anderson; Julius Louis & Brothers.

ILLINOIS.

Chicago. E. P. Peacock; Lill's Chicago Brewery; G. W. Simpson; Chicago Water Works.

Joliet. E. Porter.

Galena. John Westwick's Foundry; C. R. Perkins.

Olney. Thomas L. Luders; John Hodgson.

Salem. Hederick and Sievers.

WISCONSIN.

Beaver Dam. Chandler, Congdon & Co.'s Woolen Mill.

Richfield Station. R. R. Price.

Mendota. Wheeler & Hagel.

IOWA.

Burlington. The Burlington Gas Light Company.

NEBRASKA.

Omaha. Hall & Brother, Machinists; Richards, Crowell & Co.

MINNESOTA.

St. Anthony. Hechtman & Grethen.

St. Paul. G. W. & A. P. Merrill; Gies & Passevant.

Redwing. D. C. Hill, Manufacturer of Sashes, Doors, &c.

MISSISSIPPI.

Columbus. James Pescott.

LOUISIANA.

New Orleans. Henry Otis.

TEXAS.

Crockett, Houston County. Joseph P. Pritchard.
Waco. Joseph Giles.
Lockhart. J. G. Wiley.
Yorktown, De Witt County. C. Eckhardt.
Myersville, De Witt County. G. Hauseman.
Galveston. J. P. Davie.

CANADA EAST.

Quebec. Quebec Lunatic Asylum.

COLORADO.

Georgetown. Boston Silver Mining Association.

CANADA WEST.

Almonte. B. & W. Rosamond & Co.'s Woolen Mills, 2 boilers.
Ottawa. E. B. Eddy.

MEXICO.

Dolores Mine. A. Lowry, 3 boilers.

SOUTH AMERICA.

Lima, Peru. Stainton & White's Foundry, 3 boilers.
San Jose, Peru. Solf & Co.
Paita, Peru. George Woodhouse.
Huanchaco, Peru. Landeras, Kevia & Co.; Eugene Loyer; Alzamora, Montalvan.
Lambayeque, Peru. Tenuad & Pomar.

First Class Medal awarded at International Exhibition, in London, 1862,

"For Originality of Design and General Merit."

CERTIFICATES.

Office of William Sellers & Co.,
Philadelphia, *Aug.* 15, 1866.

Joseph Harrison, Jr., Esq.

Dear Sir,—We have your favor of the 9th inst., and may say in reply, that we have now had the *"Harrison Boiler"* in constant use in our Works for nearly two years. It has given us great satisfaction. We consider it quite as economical in the use of fuel, as any boiler we have used, or with which we are acquainted, and are satisfied that it is much ***safer than any boiler made.***

Yours, truly,

WM. SELLERS & CO.

Office of William Sellers & Co.,
Philadelphia, Dec. 27, 1866.

Joseph Harrison, Esq.

Dear Sir,—Having used your Boiler for the last two years, we are so well satisfied with its operation, that we intend ordering a second one for our new Foundry. In our experience, we have found this Boiler efficient and economical, as compared with any other with which we are acquainted, and we have subjected it to tests of great severity in our experiments upon the Giffard Injector; through all of which it has passed unharmed. It has never needed cleaning except by being blown out dry once in two weeks, and an examination, during the past summer, proved it to be free from scale. Its peculiarity in casting off and blowing out scale and deposit, under proper treatment, is most remarkable.

Yours, truly, WM. SELLERS & CO.

LINCOLN MILLS,
S. W. cor. Twenty-fifth and Spruce Sts.,
JOSEPH HARRISON, Jr., Esq. *Philadelphia, Sept.* 10, 1866.

Dear Sir,—In reply to your letter of the 9th ult., I would say that I have been using the "Harrison Boiler" for more than two years, and it gives me great pleasure to state that I find it entirely satisfactory. I have had both Cylinder and Tubular Boilers in use, and have consequently been able to compare each of them with yours. I have two of your boilers of 75 horse-power each, in use, and my engine is 70 horse-power. I do not require more than 60 lbs. of steam, but would not hesitate to run up to 250 lbs., if necessity required me to do so. I had each of the slabs tested in my presence, to 600 lbs. to the square inch. I know that it requires less fuel than the best of either the Cylinder or Tubular Boiler. My neighbor, with about the same machinery, using the steam for power generally, and heating his Mill with exhaust steam, informs me that he burns four tons of coal per day under his Cylinder Boiler, while I used less than two tons per day, during the coldest days of last winter, and heated my Mill with live steam, in addition to the amount required for power. The question of durability is one of time. I think that in consequence of the ease with which it can be cleaned or repaired, that it will last far longer than any other kind now in use. *It is perfectly safe. There is no danger whatever of explosion.* I do not hesitate to recommend it. If I ever need another boiler, I will get one of yours in preference to any other that I now have any knowledge of.

Yours, truly, SAMUEL W. CATTELL.

PHILADELPHIA ROLLING MILL,
Mr. JOSEPH HARRISON, Jr. *Kensington, Phila., Aug.* 13, 1866.

Dear Sir,—I will say in reply to yours of 9th inst., that I have had one of your Boilers almost in constant use over one of my Puddling Furnaces for over 18 months, and in all that time it required no repairs, with the exception of changing a few light bolts for heavier ones, and it is now running without any signs of leaking or want of repair, apparently as good as when first put up. I think I have just grounds, from the experience I have had, to recommend them as *a good and safe boiler*, and one that generates steam very fast. I feel confident that I get nearly double the quantity of steam from this boiler that I do from any other Puddling Furnace in my Mill that has two Cylinder Boilers over them. I believe the day is not far distant when they will be in general use in iron manufacturing establishments.

Yours, respectfully, STEPHEN ROBBINS.

Mr. JOSEPH HARRISON, Jr. *Philadelphia, Aug.* 9, 1866.

Dear Sir,—In reply to your communication respecting our opinion of the "*Harrison Boiler*," we would state as follows: We have had one of your Boilers in constant use for twenty-two (22) months, during which time it has supplied steam to a 6-horse engine, driving about 7 lathes and several other power tools. It is perfectly tight and free from leakage; takes up less room than an ordinary Boiler; and, as to its economy in fuel, you can best judge for yourself, from the following statement: During the past year, it has burned from 50 to 60 tons Pea Coal; (each week averaging 6½ to 7 days.) We can truly recommend said Boiler, from our own experience, *as safe*, reliable and economical.

Truly yours, TAWS & HARTMAN,
1237 N. Front St.

Mr. JOSEPH HARRISON, Jr. ARTISAN HALL, 611 & 613 Sansom St., Phila.

Dear Sir,—We take great pleasure in testifying to the merits of your Boiler, as a generator of steam, the confidence we have *in its safety*, its economy of fuel, and also of space for its erection. It has now been in successful operation more than a year, without the necessity of any repairs, and our confidence increases with its use. We shall always consider it a privilege to exhibit and explain its merits to any who may wish to examine it.

Respectfully, &c., GEO. W. SIMONS, BRO. & CO.

Mercantile Printing Rooms, Franklin Building,

Jos. Harrison, Jr., Esq. *Philadelphia, Aug.* 16, 1866.

Dear Sir,—I am very much pleased with the Boiler you put in for me some nine or ten months ago. It has been in constant use—no trouble—no repairs—no stopping to clean out, and steam can be "got up" in about twenty minutes. It requires less coal than the Cylinder Boiler formerly used here, although it is doing a great deal more work. I cheerfully recommend it as being and doing all that you claim for it.

Yours, very respectfully, JAS. B. RODGERS.

Jos. Harrison, Jr., Esq. *Philadelphia, Sept.* 10, 1866.

Dear Sir,—Your communication of the 9th ultimo, came duly to hand, in which you ask my opinion as to the safety, economy in fuel, and general merit of the "Harrison Boiler," based upon my experience in its use. In reply, I would state that my Boiler has been in use about nine months, and is now working in a very satisfactory manner. *I am convinced of its entire safety;* the results in the consumption of fuel are very favorable in comparison with other Boilers I have had in use. Upon the whole, its merits are very decided in many respects, and under careful management it will, I believe, be very durable. If I required additional steam-power, I would give your Boiler the preference over the wrought iron Boilers.

Yours, truly, HENRY BOWER.

Mr. Joseph Harrison, Jr. *Philadelphia, Sept.* 20, 1866.

Dear Sir,—It affords me pleasure to state that your Boiler has proved, in every particular, entirely satisfactory. As a generator of steam, it certainly cannot be equalled, either for economy, *safety* or reliability. I think we test it fully when we boil by steam 2,000 lbs. of sugar per day, run an engine, and heat four stories of the building, and all with one ton of coal per week. (The Boiler being 10 horse-power.) I have and shall recommend it, knowing its merits.

Very respectfully, J. H. SCHENCK, 15 N. Sixth St.

Daily Evening Bulletin,

Joseph Harrison, Jr., Esq. *Philadelphia, Sept.* 1, 1866.

Dear Sir,—We have had one of your 31 horse-power Globular, Five-Slabbed Boilers, known as the "Harrison Boiler," in use now nearly five months, and as a *safe, reliable steam-boiler,* and for economy of fuel, we think it cannot be equalled.

We have a 10 horse-power engine, running eight hours per day, with an average saving of 50 per cent. in the use of fuel, over the old style boiler. Our engineer, Mr. George Lodge, has had over thirty years' experience in the management of boilers, and he has no hesitation in pronouncing the Harrison Boiler the "*Best*" he ever worked.

Very respectfully yours,

EVENING BULLETIN ASSOCIATION,

607 Chestnut Street.

Joseph Harrison, Jr., Esq. *Philadelphia, August* 15, 1866.

Dear Sir,—Before ordering one of your Boilers, we sought information respecting them, from several of our friends who were using them. Their testimony was of such a character that we felt no hesitation in adopting it, and it has more than answered our expectations. We recommend them as *safe,* very economical, and easily managed; they possess *fully* all the advantages you claim for them.

Very respectfully yours, L. MARTIN & CO.,

Manufacturing Chemists, City Office, 140 South Wharves.

Removed to North Eleventh and Third St., Williamsburgh, N. Y.

JAMES, KENT, SANTEE & Co.,
Importers and Jobbers of Dry Goods,
Nos. 239 and 241 North Third Street,
Philadelphia, October 6, 1866.

Mr. JOSEPH HARRISON, Jr.

Dear Sir,—Permit us to congratulate you upon the great success of your new Steam Boiler. We have now had it in use for the last three months, and are much pleased with it.

Ours is one of your 10 horse-power Boilers, with which we run a 6 horse-power engine, pumping and compressing air enough to supply three air hoists in constant use, and besides, we have attached to it one of "James P. Wood & Co.'s Steam Heating Apparatus," through which it heats *complete*, our entire stores and counting-houses, in all containing a floor area of one acre and three eighths, with ceilings twelve to fifteen feet high. Before our store was destroyed by fire (last spring), they were heated very *imperfectly* by large heaters, while now they are heated *complete* in every department.

Your Boiler, for economy, *safety*, simplicity in its management, etc., is everything that could be desired for steam-power, and, in addition, we think one of its great merits is for heating purposes, as it takes but fifteen to twenty minutes to get up steam, so that our building is comfortably heated very early in the morning.

We think the cost of fuel now, for running the engine, and heating our stores complete, is not quite *half* as much as it cost us before, for heating alone, and very imperfect heating at that.

With our best wishes for your success in this valuable invention,

We remain, very truly yours,
JAMES, KENT, SANTEE & CO.

PENNSYLVANIA HOSPITAL FOR THE INSANE,
Philadelphia, August 11, 1866.

My Dear Sir,—In my Annual Report of this Institution, for 1865, I stated my high estimate of your Boiler, *for safety*, economy and general efficiency. Additional experience has tended to confirm all that I then said, and if we required additional Boilers, for any purpose, I should certainly recommend yours.

Very truly yours, THOMAS S. KIRKBRIDE.

JOSEPH HARRISON, Jr., Esq., Philada.

Jos. HARRISON, Jr., Esq. *Philadelphia, Sept.* 28, 1866.

The Boiler I bought of you, I have had in use for nearly three years, and it has given me entire satisfaction. *I consider it perfectly safe,* and as economical in fuel as any other boiler. Yours,

HENRY GEHRING.

JOSEPH HARRISON, Jr., Esq. *Philadelphia, August* 10, 1866.

Dear Sir,—Having charge (as administrators) of the Worsted Mills of the late Mr. Samuel Yewdall, at which the recent terrible explosion of a wrought iron boiler occurred, we have decided to avoid a recurrence of such a calamity in the future; and believing your boiler to be the only one *absolutely free from danger from explosion,* and at the same time equal, if not superior, as a generator of steam, and in economy of fuel, to any boiler now in use, you will please accept our order to furnish us for said Mills, two 50 horse-power boilers, to be used separately or in conjunction. By complying quickly with the above order, you will very much oblige

Yours, truly,
JAMES HUNTER, } Administrators.
N. R. SUPLEE, }

Mr. JOSEPH HARRISON, Jr. *Germantown, August* 16, 1866.

Dear Sir,—About four months ago, we put in one of your "Harrison Boilers," and it gives us much pleasure to be able to state that, as a *safe steam-generator,* in its general economy in fuel, time, etc., we consider it the best boiler now in use. Our boiler is 50 horse-power; our engine is a 10-inch cylinder, with a 36-inch stroke, making 50 revolutions: the cost of running this, and almost always at its utmost capacity, is about $2 per day. In fact, we consider your Boiler so excellent in its services, aside from its safeness from explosion and its real economy, that we could not, and would not do without it. It will afford us much pleasure to show the "Harrison Boiler" to any one who may call at our Works, where they can daily see it in practical operation.

Very truly yours, etc., SELSOR, COOK & CO.,
Manufacturers of Edge Tools, Hammers, etc.,
Armat St., Germantown, Philadelphia.

MOORHEAD CLAY WORKS,
Mr. JOSEPH HARRISON, Jr. *Spring Mills, Pa., Dec.* 28, 1866.

Dear Sir,—I have now thoroughly tested the Harrison Boiler you sold me, and am fully convinced that it has all the merits you claim. *That it is safe*, is beyond doubt in my mind. My engineer has raised 40 lbs. of steam in 20 minutes from the time of lighting, after the stack has been three days cold. It has not leaked any whatever, since it was put up; and in economy in fuel, it is superior, in my opinion, to any other boiler brought to my notice.

Very respectfully yours, WM. L. WILSON.

OFFICE OF THE SALEM COAL COMPANY,
JOSEPH HARRISON, Jr., Esq. *Philadelphia, August* 16, 1866.

Dear Sir,—After having your cast iron boiler in use at the Colliery of this Company for more than a year, it gives me pleasure to state that its operation has been very satisfactory. In the important point of economy of fuel, it is reported to be superior to any other boiler we have in use, and, *as regards its safety from destructive explosion*, it certainly has no equal among all the various forms of boilers that have come under my notice.

Very truly, JNO. C. CRESSON, Pres't.

ALPINE MILLS, Howards, Centre Co., Pa.,
JOSEPH HARRISON, Jr., Esq. *September* 8, 1866.

Dear Sir,—It gives me great pleasure to be able to inform you that your Boiler comes up to the most sanguine expectations; in fact, all that you can possibly claim for it: being economical, *safe* and a speedy generator of steam. Since they were first put up in the spring, (which, by the way, was done without having a mechanic on the ground, except the mason,) according to your plans, sent gratis, the first leak, trouble or delay has yet to make its appearance. Steam is kept up from 75 to 90 lbs., for Wm. H. King's (1015 Sansom Street) 25 horse-power oscillating engine, with saw-dust; there being but a 25 feet iron stack of two feet diameter. * * * * * *

I am, dear sir yours, very respectfully,
PERCY H. WHITE, Agent.

PENNSYLVANIA STATE LUNATIC HOSPITAL,
JOSEPH HARRISON, Jr., Esq. *Harrisburg, Sept.* 17, 1866.

Dear Sir,—We have had your Boiler in constant use for three months, and can say that it is all that you promised it would be,—a rapid generator of steam, and economical in the amount of fuel used. With it we can obtain the steam needed for all the purposes for which it is designed, with *half* the amount of coal we were obliged to use before, with no greater degree of pressure on the steam-gauge.

Very respectfully yours, JOHN CURWEN, Sup't.

Mr. Joseph Harrison, Jr. *Adamstown, Pa., Sept.* 10, 1866.

Dear Sir,—We are highly pleased with the Boiler we got from you last Spring. We have tested it perfectly, and think it is the best Boiler now in use, for getting up steam quickly, and for saving fuel particularly. * * * * We cheerfully recommend it as being all that you recommend it or claim.

Respectfully yours, HENRY STAUFFER.

Superintendent's Office, Camden and Atlantic R. R.,

Joseph Harrison, Jr. *Camden, N. J., August* 21, 1866.

Dear Sir,—You ask our opinion of the safety, economy in fuel, and general merit of the Harrison Boiler we have in use. I deem it a safe Boiler: from its construction, *I do not think it possible that a disastrous explosion can occur.* It is a rapid generator of steam, and requires less fuel than any Boiler that has come under my notice.

Very respectfully yours, G. W. N. CUSTIS, Sup't.

Atlantic Mills,

Mr. Joseph Harrison, Jr. *Elwood, Atlantic Co., N. J., Aug.* 13, 1866.

Dear Sir,—We have had one of your Six Slab Boilers in use in our Paper Mill for five months. We consider it unequalled by any other make of boiler now in use. With less than one-half the fuel, it produces more and drier steam than any boiler we ever used. It is simple, easily managed, *and perfectly safe.* Our Boiler bleaches the stock for, and dries one ton of paper daily, with one cord of pine wood per day.

Very truly, McNEIL, IRVING & RICH.

Joseph Harrison, Jr., Esq. *New York, October* 15, 1866.

Dear Sir,—Having thoroughly tested the merits of your excellent Steam Boiler, I have no hesitation in saying, that, in my opinion, it is the very best Boiler now in use. My engineer states that he obtains at least one-third more power from the same amount of fuel consumed, than from any other boiler he ever used.

This fact, together with its *entire safety* from explosion, to say nothing of its many other superior qualities, will certainly (as soon as understood) cause it to supersede all other steam-generators. With high regard for your enterprise, in giving to the world so valuable an invention,

I remain yours, truly, T. S. C. LOWE.

Mr. Joseph Harrison, Jr., Phila., Pa. *New York, August* 15, 1866.

Dear Sir,—We take pleasure in informing you that the Boiler purchased from you, which we have had in use about five months, has given the best satisfaction, and has borne out everything you claimed for it. As a steam-generator, we have never seen anything equal to it. We consider the saving of fuel as being very great as compared to ordinary boilers. If we had need of more steam capacity, we should most certainly use your boiler in preference to any other. You are at liberty to use this, if it will be of any service to you.

Yours, truly, United States Watch Co., F. A. GILES, Pres't.

Rock Island Manufacturing Co.,

Mr. Joseph Harrison, Jr. *Charlotte, N. C., August* 23, 1866.

Dear Sir,—Our experience with your Boiler warrants us in bearing testimony to its superiority over any other with which we are acquainted. Ours is a 100 horse-power boiler, and drives six sets of woolen machinery, and furnishes steam for our dyeing operations, and for heating the mill. Our fuel is wood, and we use three cords per day to do all our work, whereas we formerly used that quantity under Cylinder Boilers, merely to furnish steam for our dye-house, and heating the mill. Our experience is, that in fifteen minutes after applying the fire in the morning, we have on a full head of steam, and our machinery at work. We have had it in use only a few months, it is true, but we presume long enough to test its adaptation to our fuel and our work, and have found it, in every respect, to come up to your representations. Our Boiler was set up and put to work by a man who never had seen it done, without the slightest difficulty. Your Boiler commends itself for economy in fuel, and its merits need only be known to render it universally popular.

Very respectfully, yours, JOHN A. YOUNG, Pres't.

J. Harrison, Jr., Esq., Phila. *Boston, October* 23, 1866.

Dear Sir,—I have just received an order from Messrs. Stainton & White, Lima, Peru, South America, for one of your twenty-five (25) horse Harrison Boilers, to be shipped at the earliest possible moment, with all the fittings the same as the 10 horse-power Boiler shipped to them in March last. They have had the Boiler in constant use in their shop since it was received, and it gives great satisfaction. All who have seen it are delighted with the idea of *a boiler which cannot be blown up. The 25 horse-power Boiler is to replace a Cylinder Boiler which blew up lately.*

Yours, respectfully, GEO. W. WHITE,
48 & 50 Union Street, Boston, Mass.

Richmond Rolling Mill,
Mr. Joseph Harrison, Jr. *Philadelphia, May* 17, 1867.

Sir,—We have had one of your boilers in constant use for nearly eighteen months, and during that time nothing has been done to it but replacing two or three defective balls, which took but a few hours to do.

We cheerfully testify to its superiority over all boilers now extant, in these respects, viz.: speedy generation of steam; perfect safety from explosion; economy of fuel; easy access to all parts for repairs or cleaning. Combining these qualities with the small space occupied by them, we incline to the opinion that should we have occasion to require more boilers, or to replace the cylinder boilers also in use, we would prefer the "Harrison."

Truly yours, &c.,
(Signed) DU PLAINE, COOPER & STOKES.

Keystone Saw Works, Nos. 67 and 69 Laurel St.,
Mr. Joseph Harrison, Jr. *Philadelphia, May* 31, 1867.

Dear Sir,—I am in receipt of yours of the 30th inst., and in reply, state, that for the purpose I am using your boilers, that is, to take up the heat over my heating furnaces, they are "just the thing." I would not be without them, and I do not know what I could substitute for them.

We cannot burst them, for that we have tried; by accident we got the water down with a hot fire still under them, got the balls red hot, and all we had to do the next morning was to screw them up, and off they went again as good as ever. Any other boiler would either have blown up or been ruined by the heat, but this is as perfect as ever. I think they are the best natured boilers in the world.

Respectfully yours,
(Signed) HENRY DISSTON.

Mr. Joseph Harrison, Jr. *Philadelphia, June 11th,* 1867.

Dear Sir,—Having tried one of your Boilers for five or six months, I am very well pleased with it. For safety and economy of fuel it is better than any I have had before. I don't know any boiler that I would exchange for it.

Yours truly, DAVID FRANCE.

JOSEPH HARRISON, Jr., Esq. *Philadelphia, June 4*, 1867.

Dear Sir,—In response to your letter we would say, that we have had a 16 horse-power boiler of your make, in use in our mill since last August, and it has given us great satisfaction. It is easily managed, steam can be got up quickly, and we believe it is more economical in the use of fuel, than the tubular boiler which we formerly used.

Yours, respectfully,

(Signed) JOHN G. WHITE & CO.,

Malt Mill, No. 424 Dillwyn St.

Mr. JOSEPH HARRISON, Jr. *Philadelphia, June 4*, 1867.

Dear Sir,—We have had your boilers in operation for over four months, and can safely say they are good steam-generators, and perfectly safe. We consider them equal to any other boiler in economy of fuel. They have given us entire satisfaction. Were we putting in any more boilers, would *certainly* put the "Harrison" in preference to others. Would take great pleasure in recommending them to any one.

Yours, respectfully, THORNTON & MAYBIN,

Wamsutta Mills, Tenth and Columbia Avenue.

Mr. JOSEPH HARRISON, Jr. *Philadelphia, June 3d*, 1867.

Dear Sir,—Your letter of the 30th ult., asking our opinion of your new steam-generator, is received, and in reply we have no hesitation in saying we approve of all its merits.

We formerly used a cylinder boiler, and experienced great trouble in getting up steam, thereby losing much time; and we were hesitating as to what kind of boiler to use in place of it. We finally decided to try one of your make.

We have had occasion while using some inferior coal, to draw the fire and light a new one; we did so, and during the time continued running full speed.

We think one of the great merits of your boiler is, safety from explosion; this alone should bring it into general use.

We are always willing and ready to show and explain it to those calling.

You can make any use of this you may think proper.

Yours, most respectfully, WILBANK & HOWARD,

Shear-Blade Manufacturers and Machinists.

Nos. 2112 and 2114 Filbert St.

JOSEPH HARRISON, Jr., Esq. *Providence, R. I., June 7th*, 1867.

Dear Sir,—The best evidence we can give you of our satisfaction with the 16 horse-power boiler we bought of you a few months since, is to order another precisely like it, which we now do, and should like to have it forwarded with the same fittings and extra appurtenances as the last.

We should like to receive the above at your earliest convenience, by outside steamer from Philadelphia to Providence.

Your obedient servants, T. P. SHEPARD & CO.

Wilmington, Del., June 12th, 1867.

Mr. JOSEPH HARRISON, Jr., Proprietor Harrison Boiler Works.

Dear Sir,—We have been using one of the "Harrison Boilers" for nearly a year, and have no hesitation in saying, that, in our opinion, it is far superior to any other for safety, economy of fuel, &c., and it affords us much pleasure to recommend it to those in want of good, substantial and economical boilers.

We are, dear sir, very respectfully,

HARTMANN & FEHRENBACH, Brewers, Wilmington, Del.

JOSEPH HARRISON, Jr. *Fairhaven, Vt., June 8th,* 1867.

Dear Sir,—We received your inquiries asking how we liked your boiler, &c. In reply will say that we are much pleased with it; and as far as our experience goes, we believe it is all you recommend, and think we make a saving of at least one-third of the fuel over our old boiler, while it does our work much better. As to its safety, we made a good test some three months ago. The fireman went off duty, leaving the water low, and the next man came on late; seeing the water low and steam also down, he thought he would get up steam first, and then put on the pump; but seeing the steam did not come up, we looked into the furnace and saw it was at a white heat; as it was very hot, we let it cool off and then put on the pump, but the latter leaked out as fast as it came in. We then screwed up the nuts on the bolts, and found the boiler as good as ever. It has been at work ever since, and works just as well as it did when first put in.

Respectfully, yours, T. & W. MILLER.

JOSEPH HARRISON, Jr., Esq. *Lynn, Mass., June 11th,* 1867.

Dear Sir,—In reply to yours of the 30th of May, we would say that your Boiler has come up to our fullest expectations. As to its general merits, it cannot be surpassed by any boiler ever made. It has never leaked a drop, and we feel safe at any pressure under which we ordinarily run.

Last week, through the carelessness of the engineer, (in over-weighting the safety-valve), the gauge indicated 160 pounds.

Our engineer works in the shop seven hours out of the ten.

As far as economy of fuel is concerned it is unquestionable.

Yours respectfully, HOLBROOK & GLAZIER.

Peru, Ill., June 25th, 1867.

JOSEPH HARRISON, Jr., Esq., Harrison Boiler Works, Phila.

Dear Sir,—Your favor of the 30th ult., asking my opinion of the 50 horse-power Boiler I bought of you, came safely to hand.

Allow me to say that I am very much pleased with the Boiler. I (who never saw a boiler set) superintended its erection, and a young smart fellow, not 21 years of age, coupled it to the engine, and it has now its complement of water, and does not leak a drop. We had 70 pounds of steam in 30 minutes, and the furnace had no fire in it for two weeks.

Allow me also to express my great admiration of the excellence of the workmanship in your Boiler and fittings, and the very reasonable prices at which they are furnished.

Yours truly, JAMES BARTON.

JOSEPH HARRISON, Jr., Esq., Phila. *Portland, Me., July 19th,* 1867.

Dear Sir,—After several months constant use of the "Harrison Boiler," it gives us great pleasure to state that we consider its good qualities cannot be equaled by any other boiler we have ever seen.

From some severe tests, both before and after setting in operation, we are confident as to its entire safety from explosion, and we can cordially endorse all claims as to its facility for generating steam, and also the readiness with which it can be cleaned inside and out.

In economy of fuel it exceeds our most sanguine expectations, and we could make some statements thereto that might appear incredible where we are not known. We therefore prefer and should take pleasure in exhibiting to parties interested a practical illustration of its superiority in this and other respects.

To parties wanting steam, particularly for boiling or heating purposes, from our experience, we consider your Boiler peculiarly adapted, and cannot recommend it too highly.

Yours truly, LEATHE & GORE,
Manufacturers of Steam Refined Soaps.

Camden, N. J., December 20, 1867.

Mr. Joseph Harrison, Jr., Proprietor Harrison Boiler Works.

In reply to yours under date of December 20th, 1867, regarding the "Harrison Boiler," we would say, that having used one for a year, we feel justified in saying, that for economy of fuel, safety, quick generation of steam, and other general advantages, with far less liability to become out of order, obtained by the use of the "Harrison Boiler," we consider it a very profitable investment.

Very respectfully, yours, GARRISON, GILLINGHAM & CO.

Mr. Joseph Harrison, Jr., *Lebanon Furnace, December* 21, 1867.

Dear Sir,—It gives me great pleasure to state that the boiler you furnished me for my repair shop has given me entire satisfaction.

I can raise steam in a shorter time than in any boiler I have ever used.

It has not leaked a drop since it was put up and has not cost a cent for repairs. I feel that it is a very safe boiler, as I do not see how a dangerous explosion can take place.

I believe it to be an economical boiler, though I have made no experiments with it as to the amount of coal consumed in work done.

Yours, truly, G. DAWSON COLEMAN.

Office Nashua Lock Co.

Joseph Harrison, Jr. *Nashua, N. H., December* 23, 1867.

Dear Sir,—We have had one of your "Harrison Boilers" in our establishment since March last, and can say that for "safety" nothing can surpass it. As you know, one section was dropped into the hold of the vessel in shipping it, which has caused two of the balls to break both times when our steam was at 80 pounds, and no damage done. As to the economy of the boiler, we cannot say too much in its favor, and as to the generating of steam, nothing, we think, was ever made to surpass it. Our boiler is 50-horse; we use 32 of it for our engine, and from the top of the boiler a 1½-inch pipe starts and goes through 1800 feet of same sized pipe into a Japan oven, and the steam passing continually through it, and it never lacks for steam. We cannot speak too highly of all its merits.

Yours, truly, F. O. MUNROE,
Superintendent N. L. Co.

Joseph Harrison, Jr. *New Britain, Conn., December* 24, 1867.

Dear Sir,—Your letter was duly received. In reply will say we are very much pleased with the boiler we had of you last summer. We find it to be all you claim for it. We are convinced of its entire safety, and as for economy, we want no better proof than our engineer asking us for an increase of his wages, as he said he only used about one-half of the fuel required for the other kind of boilers to keep up the same amount of steam. We know of no other boiler that steam can be got up as quickly, with as little attention, and what we now know of the Harrison Boiler, we should prefer it to any other we have ever seen.

Very respectfully, yours, JUDD & BLAKESLEE.

Office of Van Brunt & Co., Manufacturers of Agricultural Implements,

Mr. Joseph Harrison, Jr., Phila. *Horicon, Wis., December* 25, 1867.

Dear Sir,—Yours of the 20th inst. making inquiries as to the working of the Harrison Boiler had of you, we have to say that we have been well pleased with its power to generate steam, exceeding any other boiler we have ever used in economy of fuel and facilities of getting up steam.

Yours, respectfully,
VAN BRUNT & CO.

Almonte, C. W., December 26, 1867.

Mr. Joseph Harrison, Jr., Harrison Boiler Works, Phila.

Dear Sir,—In reply to yours of the 20th it gives us great pleasure to be able to bear testimony in favor of your Boiler on all points on which you ask our opinion, safety, economy in fuel and general merit. We have now two of 75 horse-power each, and are very much pleased indeed that we did not put in tubulars, as we had at first intended. In reply to a letter from another Canadian firm, we wrote this morning: "A most comfortable thing is to be able to stand in front of a boiler and have no feeling of danger." Such is our experience, and we have no hesitation in recommending your boiler to all in want of Steam-power.

Yours, truly, B. & W. ROSAMOND & Co.,
Manufacturers of Woolens.

Black Diamond Steel Works, Park, Brother & Co.
Pittsburgh, Pa., December 26, 1867.

Joseph Harrison, Jr., Esq., Harrison Boiler Works, Phila.

Dear Sir,—For some five months we have used your boilers over two of our heating furnaces, using the steam from them to operate one of our three ton hammers.

The boilers will make more steam than we require, and consequently we have our heating furnaces with spare or additional stacks, so arranged that when the heat from the furnaces gives us more steam than wanted, we close the dampers of the furnaces on the rear stacks and open those belonging to the forward stacks, thus preventing the waste heat from the furnaces from passing under and along the boilers.

We believe your boilers are less liable to accident than any kind of steam generator now in use, and we would not substitute any other description of steam-boiler in place of them, for the following reasons, viz.:

We believe them to be the safest, most economical and durable boilers now offered to the public. Since the first day we fired up our heating furnaces, over which we have your boilers placed, we have not experienced the least trouble, and do not expect to have any trouble in the future; we therefore cheerfully recommend your boiler to the public.

Respectfully, yours, PARK, BRO. & CO.

Joseph Harrison, Jr., Esq., *Providence, R. I. December 26, 1867.*

Sir,—We have had in operation one of your boilers of one hundred and fifty horse-power since the first of June, 1867, and it is but justice to say that we have found it to come fully up to your recommendation.

We find it economical in fuel, and are satisfied as to its perfect safety.

It has given us no trouble from the start, is easily managed and is all we could desire. To show our perfect confidence in it, we have put in a second boiler of the same description and power, which is in successful operation.

Respectfully, yours, ALBERT ARMINGTON,
Superintendent of Silver Spring Bleachery.

Office of the Galena Foundry,

Joseph Harrison, Jr., Esq. *Galena, Ills., Jan. 6, 1868.*

Dear Sir,—It is with pleasure that I bear testimony to the merits of the Harrison Boiler. I have had your boiler in use about one year, and it has given entire satisfaction. It has had the most severe tests, which would have ruined any ordinary boiler. It is very economical, and takes very little fuel to raise steam. We have raised steam to 40 pounds in *eight minutes*, on Monday morning, when the water was cold. In thirty years' experience with boilers in England and America, I have not found one to compare with the Harrison Boiler for safety and economy.

Yours truly,
JOHN WESTWICK.

Office of Hance, Griffith & Co., Manufacturing Chemists and Pharmaceutists.

Mr. Joseph Harrison, Jr. *Philadelphia, December* 27, 1867.

Dear Sir,—In answer to yours of the 20th inst. it gives us pleasure to state that we are well satisfied with our "Harrison Boiler," deeming it economical in fuel, easily managed and have confidence in its exemption from destructive explosion.

During the year it has been in use, our works have not been delayed one hour, nor have we discovered the least imperfection either from leakage or any other cause.

Having taken the precaution to secure sufficient capacity, we have ample steam and considerable to spare, burning a slow, moderate fire, with the steam damped (set at 45 pounds) closed most of the time. Very respectfully, your friends,

HANCE, GRIFFITH & CO.

Jos. Harrison, Jr., Esq., Phila. *Alexandria, Va., December* 26, 1867.

Sir,—We have had your boiler in use about eight (8) months, and find that in economy of fuel, safety and superior finish, it is all that was promised, and we are entirely satisfied with it. Very respectfully, yours, etc.,

J. G. VERPLANCK & CO.

Joseph Harrison, Jr., Esq., *Neponset, Mass., December* 27, 1867.

Dear Sir,—We purchased from you the early part of the year one of your "Harrison Boilers," and from the experience we have had with it so far, we would say, that were we about to purchase any more, we should most decidedly give yours the preference.

Yours, truly, S. S. PUTNAM & CO.,

Manufacturers of Forged Horse Nails, &c.

Joseph Harrison, Jr., Esq. *Loyal, Clark Co., Wisc., Jan.* 8, 1868.

Dear Sir,—I am well pleased with the Harrison Boilers. I believe them to be perfectly safe, and I know they cannot be equalled for economy in fuel. My pump got out of order and we ran some time. We had no gauge of water, and did not know where the water was, it must have been very low. When we had pumped some time, the boiler began to leak quite badly, so I put on the wrench and gave a partial turn, and in about ten minutes it was tight again, and remained so. We run with 110 lbs. steam. Yours truly,

JOHN GRAVES.

Mr. Joseph Harrison, Jr. *Philadelphia, January* 9, 1868.

Dear Sir,—I have received your favor of 20th ult., and am happy to state that your boiler is perfectly satisfactory in all respects. If it only continues to work as well as it does at present, it seems to me to be the perfection of things. I believe that your boiler is rather better in point of economy than the locomotive tubular boiler I am also using, of which I had great doubts when purchasing yours. It has also a decided advantage over all other boilers that I have been acquainted with in the dryness of its steam, which to me is of considerable advantage, using it as I do principally for boiling and drying purposes. There is another peculiarity of your boiler, which I have been unable heretofore to find in any other, viz.: that of burning wet fuel, such as tan, spent dye-wood, etc. It does this in a most satisfactory manner, without that abundance of soot and black smoke which made the burning of it heretofore as much a nuisance as it was to find a way to get rid of it otherwise. Previous to the fire at my place, when we were not entirely dependent upon one boiler, I have driven a twelve-horse engine, and done considerable of the boiling and drying by the sole use of this material—spent dye-wood, effecting a great saving and getting rid of a great nuisance.

In conclusion, I have a forty-five horse-power boiler, locomotive tubular pattern, for which I was, possibly a year ago, offered $2,500. It is in perfect order, and I will exchange with you for a fifty-horse, similar to the one I have, and give you $500 cash besides.

I remain, very respectfully, DANIEL ALLEN.

PHŒNIX STEAM MILLS.

JOSEPH HARRISON, Jr., Esq. *Philadelphia, January* 10, 1868.

Dear Sir,—I am in receipt of your favor asking my opinion of the "Harrison Boilers," put up in the place of the boilers that exploded with me in January last, and which have been in constant use ever since. I find they save me at least 25 per cent. in fuel, and less liable to leak, and I consider them entirely safe, and prefer them, in every respect, to the old-fashioned boiler. Yours, very respectfully,

JAMES M. PATTON,
Brown Street Wharf.

CINCINNATI TYPE FOUNDRY,
No. 201 Vine Street,

JOSEPH HARRISON, Jr. *Cincinnati, Ohio, Jan.* 21, 1868.

Dear Sir,—We have had your boiler in use seven months, giving us great satisfaction. We are using the same amount of fuel we did under our old double-flued boiler, and are running our shafting twenty-five per cent. faster, having plenty of power, whereas we formerly suffered from a lack of it. We regard it as a very safe, reliable and economical boiler. Yours, &c., C. WELLS,

Treasurer Cincinnati Type Foundry Company.

JOSEPH HARRISON, Jr., Esq. *Newburyport, Mass., Jan.* 21, 1868.

Dear Sir,—We will say, in reply to yours of the 20th December last, that we have had one of your boilers (50 H. P.) in use for nearly a year, and it has given perfect satisfaction in every respect. It is economical in the use of fuel, and, in our opinion, is perfectly safe from explosion through carelessness or accident.

Very truly yours, C. H. BALLARD,
Superintendent Merrimac Arms and Manufacturing Co.

MR. JOSEPH HARRISON, Jr. *Springfield, Ohio, Jan.* 22, 1868.

Dear Sir,—Your circular letter of the 20th ult. was duly received. Owing to ill-health, have not been able to answer sooner. The boiler I purchased of you I have had in operation about eight months. As you may be aware, I do not run an engine; but use the boiler for coloring, scouring, drying, heating, &c., &c.; consequently cannot say as to the capacity of the boiler. As to its durability, safety, speed in getting up steam, I am entirely satisfied. My own opinion is, that perhaps it will not generate steam as fast as the modern tubular boiler, but much faster than the flue boiler. I am inclined to think, taking it altogether, it is the *best boiler* in use to my knowledge. Yours truly,

CHARLES RABBITTS.

JOSEPH HARRISON, Jr., Esq. *Danielsville, Pa., Jan.* 27, 1868.

Dear Sir,—In reply to yours relative to our opinion of the twenty-horse power Harrison Boiler in use at our quarry for the year past, we must say that its performance is, in every way, most satisfactory. As a rapid and economical steam generator it cannot be excelled, and for safety from explosion it deserves the highest recommendation. Our experience with it entitles it to preference over all other boilers.

Yours respectfully, MORRIS PUGH,
Superintendent Mauch Chunk Slate Co.

JOSEPH HARRISON, Jr. *Providence, Feb*: 1, 1868.

Sir,—It gives me great pleasure to recommend the "Harrison Boiler" for general use, both for economy and general safety. For the quick generation of steam we think it unequalled. Most respectfully yours,

A. A. DRAPER, Agent,
National Brick Company.

MOOREHEAD CLAY WORKS,
Spring Mill, Penna., Feb. 8, 1868.

MR. JOSEPH HARRISON, Jr.

Dear Sir,—Since we repaired the balls which leaked in the "Harrison Boiler" we procured of you, it has continued to give us entire satisfaction. It is but just to say that the leak was no fault of the boiler; but was in consequence of our engineer meeting with an accident, and an oversight of those left suddenly in charge of the engine-room, leaving the water to become exhausted, and then at once turning in fresh. When we reflect that this mistake, which was fully repaired in three hours, without interfering with our work, might, had we been using an ordinary boiler, resulted in another of those disasters, of late too frequent, we cannot but send you this voluntary tribute of gratitude for the great safeguard which your enterprise has thrown around the lives and limbs of our workmen, and the entire security furnished our property.

The report was prevalent here that our "Harrison Boiler" obliged us frequently to stop. We wish to say *now*, that from the time fire was first applied to our boiler, up to the present moment, we have never had the slightest cause to cease operations proceeding from the boiler. It has been a faithful and over-worked servant. Our unceasing work is ample proof that the boiler has done its full duty.

Very respectfully yours, WILLIAM L. WILSON.

MR. JOSEPH HARRISON, Jr., Phila. *Olney, Ills., Feb.* 19, 1868.

Dear Sir,—I have had many years experience in generating steam in wrought-iron boilers of various forms, but have never found any one of them safe or economical when compared with the "Harrison Boiler," which I have been using over five years past. I find it fifty per cent. more economical than any of my former boilers, and perfectly free from danger or possibility of destructive explosion. I have tested numbers of the units up to 1500 lbs. to the square inch with cold water, and am fully convinced that there is no other boiler capable of generating steam for practicable regular duty, as high in pressure or as dry in delivery, or that can at all compare with it in its marked features of absolute safety, and great economy and durability.

Yours respectfully, THOMAS L. LUDERS.

UNION SUGAR REFINERY,
Charlestown, Mass., Feb. 20, 1868.

JOSEPH HARRISON, Jr., Esq., Phila.

Dear Sir,—Your six fifty-horse power steam boilers which we have in operation since August last have given us entire satisfaction. In regard to their safety, economy of fuel and generation of steam, they surpass by far the tubular and flue boilers we had in use formerly. After we tested each of your boilers with 150 pounds steam pressure, we disconnected our old ones, and have worked our refinery since with yours only, day and night, under a pressure of 60 pounds, and never discovered any leakage in them. Four of the "Harrison Boilers" will now give us all the steam we want, at a regular pressure—a result we could never reach with our six old ones. However, we use them all six, as we find a saving in fuel and less depreciation in boilers and grates. We have used your boilers too short a time in order to give you the exact saving in fuel, and to express our opinion in regard to their durability, but will do so with pleasure at a future time. Suffice it to say that we are now convinced that they surpass by far our old, and any other kind of boiler we are acquainted with.

Yours truly, GUSTAVUS A. JASPER,
Superintendent.

MR. JOSEPH HARRISON, Jr. *Philadelphia, May* 18, 1868.

Dear Sir,—In reply to your circular letter of inquiry, asking our opinion of the Steam Boiler erected by you at our Works in 1867, we would state—

First.—For safety sake, we would not have in its place any other boiler with which we are acquainted.

Second.—We believe it to be as economical as any other boiler we have used under the same circumstances, although on this point we have made no absolute test beyond the following observation. On the 1st of February last, when the pressure of steam was seven and a half pounds, indicating a temperature of the water of two hundred and thirty-five degrees, the temperature of the gas escaping up the chimney was two hundred and ninety-five degrees, showing a difference of but sixty degrees.

Third.—Like all boilers, carrying comparatively a small stock of water, it has the advantage of raising steam rapidly, and the disadvantage (if it be one) of requiring constant watchfulness on the part of the man in charge.

The penalty for his carelessness would probably be, that he would burn out some of the globes; but we think this to be preferred to blowing up our building and destroying half of our people. We remain,

Very truly yours,

TATHAM BROTHERS.

Yorktown, De Witt County, Texas, May 13, 1868.

JOSEPH HARRISON, Jr., Phila.

Dear Sir,—Last summer we ordered for account of Messrs. Fechner & Zedler, of this place, one of the Harrison Boilers, which was put up here and has been in use ever since, and proved, in regard to safety, economy of fuel, and durability, to be in every respect satisfactory, and equal to the merits claimed for it.

Being satisfied that, on account of the above stated advantages, the Harrison Boiler is better adapted to this country than any other, we shall not fail to recommend the same to all of our friends who wish to use steam power, in order to increase its use in this part of the country. Please send us your latest price list, and believe us,

Yours truly, C. ECKHARDT & SONS.

KEYSTONE ZINC WORKS.

Birmingham, Huntingdon County, Pennsylvania, January 25, 1869.

JOSEPH HARRISON, Jr.

Dear Sir,—In reply to yours of the 7th inst., would say it gives me great pleasure to bear testimony in favor of your boilers. We have had them in use for two years. I put them up myself, and have never seen anything of the kind until they came here, and, with the aid of your draft, I had no trouble in erecting them.

They use less coal—never get out of repair. In fact, I have no hesitancy in saying they surpass any thing in the shape of boilers. Yours, E. O. BARTLETT,

Superintendent.

UNION SUGAR REFINERY,

MR. JOSEPH HARRISON, Jr. *Charlestown, Mass., Jan.* 21, 1869.

Sir,—In my last communication to you in regard to your six fifty-horse power steam boilers, I promised to give you at a future time the exact amount of saving in fuel.

I am pleased to be able to do so now, as I kept a very close and accurate account of the fuel used from December, 1867, to December, 1868, to compare your boilers with the old ones, which we took out in September, 1867.

The amount of raw sugar we refined during that time was larger than ever before. The steam pressure was always kept over fifty pounds, and we were therefore enabled to do more work in a shorter time with the same machinery and apparatus than with our old steam boilers, in which the pressure at times could not be kept up higher than twenty to thirty pounds.

The actual saving in fuel during this time was one thousand seventy-one tons of coal.

Yours truly, GUSTAVUS A. JASPER,

Superintendent.

Joseph Harrison, Jr. *Worcester, Mass., Feb.* 1, 1869.

Dear Sir,—After using one of your thirty-one horse power boilers sixteen months, it gives me great pleasure to say that it has proved in every particular entirely satisfactory. I have had many opportunities to recommend it, which I have done with pleasure. For economy in fuel, it excels all boilers in use. I can raise steam in shorter time than in any boiler I have ever seen; and would cheerfully recommend it as the cheapest, safest, and most economical boiler ever in use. Very respectfully yours, A. N. WHEELER.

Office of the Granite State Mills,

Joseph Harrison, Jr., Esq. *Newport, N. H., Feb.* 1, 1869.

Dear Sir,—Yours of January 9th, asking our opinion of the Harrison Boiler, is at hand. We have had your boiler in use fifteen months, and it gives us perfect satisfaction in quickness of raising steam, and the cheapness of running it, as compared with the old-fashioned boilers. Our boiler does not yet leak a drop, and we do not believe it ever will,—and last, though not least, our workmen about it do not feel obliged to get their lives insured. Yours truly, COFFIN & NOURSE.

Joseph Harrison, Jr., Phila., Pa. *Concord, N. H., Feb.* 8, 1869.

Dear Sir,—Your favor of January 9th came duly to hand. The 62½ horse power boiler which we had of you in October has, from the time we commenced its use until now, given perfect satisfaction, and we can see no reason why it should not continue to do so. It is the only boiler (and we have had a variety of kinds) that suited us from the first. We save fuel, are troubled with no leakage, and feel safe in using it. We can only say, it meets more than our expectations. Truly yours,

ABBOT, DOWNING & CO.

Mr. Joseph Harrison, Jr. *Germantown, Pa., May* 8, 1869.

Dear Sir,—In reply to your inquiry as to our opinion of the "Harrison Boiler," we would say, that after three years of experience, during which time it has not cost us a single penny for repairs, we find it in as good condition as when we first started it, giving us great satisfaction. We have used the tubular, flue and plain cylinder boilers, and unhesitatingly pronounce the "Harrison" superior to either; in fact it is the best boiler in point of economy and general usefulness that we know of.

Very truly yours, &c., SELSOR, COOK & CO.,

Manufacturers of Hammers, Hatchets, Edge Tools, &c.

Pennsylvania Hospital for the Insane,

Joseph Harrison, Jr. *Philadelphia, December* 27, 1869.

Dear Sir,—It gives me pleasure to be able to say that the boiler we obtained from you, three or four years since, has always given us entire satisfaction, and we have had no trouble with it of any kind. During the last winter we relied on it almost entirely for the heating of the whole of our department for females, one of the tubular boilers, used for other purposes, when not so employed, assisting it. Judging from our experience, it seems to merit all you claim for it. THOMAS S. KIRKBRIDE, M. D.

Mr. Joseph Harrison, Jr. *Royal Oak, January* 26, 1870.

Dear Sir,—Your account of the report of the American Institute was received to-day. Your boiler is *everything* you represent it to be. In a saw-mill, perhaps the most unfavorable position for it, a larger boiler than ours would doubtless be more economical, on account of the irregular and heavy drafts made upon the steam, but we frequently saw two thousand feet of lumber with the expenditure of one half a cord of wood, (slabs.)

Yours, truly, ROBINSON & BRO.

OFFICE OF CHENEY BROTHERS SILK MANUFACTURING CO.

MR. JOHN A. COLEMAN, Boston, Mass.

South Manchester, Conn., Sept. 21, 1868.

Dear Sir.—In answer to your favor of the 16th inst., in regard to the "Harrison Boiler," would say that we have one 50 horse power in constant use for eighteen months past, and six others, 250 and 475 horse power, for eight months past, and they have so far given perfect satisfaction in every way; and as far as economy in fuel is concerned, think them preferable to either the tubular or flue boiler, and were we in want of more power should by all means give them the preference.

Very respectfully yours,

CHENEY BROTHERS.

NOTE.—See Letter from CHENEY BROTHERS, dated March 22d, 1871, on page 76.

OFFICE OF F. O. MATTHIESSEN & WEICHER, Sugar Refiners.

106 *Wall Street, New York, May* 14, 1869.

JOHN A. COLEMAN, ESQ., Boston.

Dear Sir.—In reply to your yesterday's favor, we beg to say that we did not authorize anybody to make unfavorable statements in regard to the "Harrison Boiler." On the contrary, we are so well satisfied with them that we have only lately ordered another Boiler of that kind, and shall always give them the preference.

Yours truly,

F. O. MATTHIESSEN & WEICHER.

OFFICE OF WALLACE & SONS.

Ansonia, Conn., June 17, 1870.

JOHN A. COLEMAN, Agent, 110 Broadway, New York.

In reply to your inquiry as to how we are pleased with the "Harrison Boiler," purchased of you, we would respectfully present a few facts as the result of careful experiments made and continued for weeks, for the purpose of determining the practical value of your Boilers over others we have now in use. In comparison with a first-class Tubular Boiler, 48 inches in diameter, containing ninety-six 2½ inch tubes, 14 feet long, the economy of fuel in favor of your Boiler was as follows:

Tubular Boiler, 10 hours, 24,000 lbs. water, 2,700 lbs. coal.
Harrison " 10 " 24,000 " 2,125 "

One pound of coal in the Harrison evaporating 11.29-100 pounds of water, against 8.88-100 pounds evaporated in the Tubular, by the same weight of coal in the same time. The steam from the Harrison Boiler was also much drier than that from the Tubular Boiler, a fact which we consider much in its favor. These facts having been demonstrated, speak sufficiently for themselves and for the Boiler, without further comment from us. Safety and economy are the necessary adjuncts of a steam boiler, and the man who can combine these to stand a practical demonstration is the man the people wish to see.

WALLACE & SONS.

Doylestown, Pa., July 1st, 1870.

MR. JOSEPH HARRISON, JR.

Dear Sir.—The Boiler is doing everything any one could expect, and a great deal more than I expected. One and one-quarter tons of coal will keep it running a week, and the shavings from the planing machine keep it running day in and day out, and have some left. We do not use but sixty pounds of steam to run all of our machinery—four circular saws and a jig saw, panel raiser, planing, moulding, tenoning, mortice, and four other machines of different descriptions.

Yours, truly,

C. FINNEY.

South Manchester, Conn., March 22d, 1871.

JOHN A. COLEMAN, Agent, Harrison Boiler, No. 110 Broadway, New York.

Dear Sir.—In reply to your favor of the 21st inst., inquiring our present opinion of the Harrison Boiler, we would say that we have no cause to change our minds. We now have seven Harrison Boilers, making four hundred and fifty horse power, and we are well satisfied that they are the best we know of for making steam, and most economical in fuel for the quantity of steam they furnish.

It is now over five years since the first boiler was put in, has been in constant use since, and we have had no trouble at all about leaky joints. Was tested a few weeks since by the State Inspectors, and reported tight as a drum.

We can only say that if we required more power, we should by all means give the Harrison the preference, which we consider the best recommendation we could give.

Yours, truly, CHENEY BROTHERS.

Germantown, Pa., April 26th, 1871.

MR. JOSEPH HARRISON, JR.

Dear Sir.—You ask us our opinion of your Harrison Boiler. We would say in reply, we have had one in use for the last five years and we are much pleased with it. We have used the Cylinder, Tubular and Flue boilers, but the Harrison Boiler surpasses them all in safety and economy. Although we have used it for five years we can get up steam as quickly now as when we first started to use it, and this goes to prove that the Boiler is clean inside. We can recommend it to the public, believing it to be one of the best steam generators now in use. Yours, respectfully, GEO. SELSOR & CO.

Edge Tool Manufacturers,

Armat St., Germantown.

Philadelphia, February 4th, 1871.

JOSEPH HARRISON, JR., ESQ.,

Dear Sir.—I have had your boiler in use for nearly four years. I take pleasure in saying that it has given entire satisfaction. No repairs have been required during that period, and in point of safely I believe it to be all you claim for it.

Yours respectfully, H. C. FOX,

FLINT GLASS WORKS, *Sutherland Ave.*

Portland, May 3d, 1871.

JOSEPH HARRISON, JR. ESQ.

Dear Sir.—The boiler bought of you in 1866 has been in constant use, and is apparently in as good condition as when first set up. From experience we can confirm what we have heretofore expressed as to its economy in fuel, and can now add, to its freedom from repairs and also its durability. The boiler having given us perfect satisfaction in all respects during this period we can cheerfully recommend it to all parties in want as the best boiler known.

Yours, truly, LEATHE AND GORE,

Manufacturers of Steam Refined Soap,

379 Commercial Street.

Portland, Maine.

REPORT

OF

BOILER EXPLOSIONS IN ENGLAND for 1866.

By E. B. MARTEN,

ENGINEER OF MIDLAND BOILER ASSOCIATION.

Reprinted from "ENGINEERING," for March 15th, 1867.

By the courtesy of Mr. E. B. MARTEN, the Engineer of the Midland Boiler Association, we are enabled to reprint the following illustrated portion of his report for the year 1866.

No. 1. *January 1st*, 1866. *None injured.*

This was a locomotive boiler which had been shunting, and was standing with steam up near a platform. All but the fire-box was blown away, the main portion being thrown a distance of 400 yards. The explosion took place at a longitudinal seam of the barrel. The engine had been working at 110 lbs., and had been proved by hydraulic test above that pressure a short time before. The cause was grooving at the joint which gave way, which it was considered had gone on very rapidly, as it was not discovered when examined and tested.

No. 2. *January 2d*, 1866. 2 *injured.*

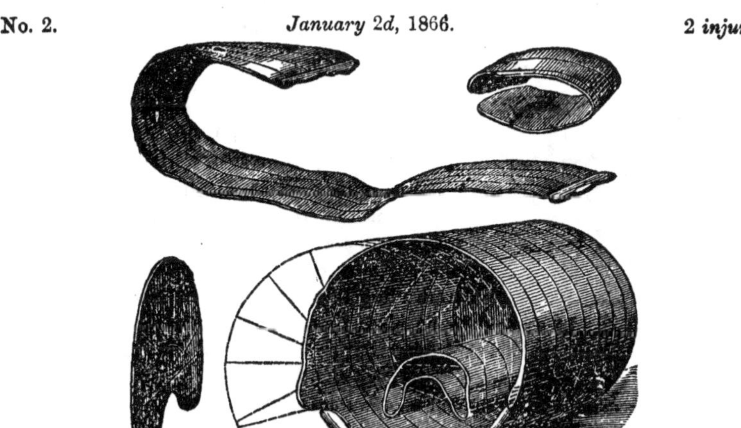

The boiler was of the type known as Butterley. It was 26 feet 6 inches long and 9 feet diameter. The wagon-shaped top to the fireplace was 8 feet 6 inches long, and was attached to the bell-mouth of the internal tube, which then continued circular to the back of the boiler. The tube was 3 feet 6 inches diameter. All the plates were about $\frac{7}{16}$ inch thick, and although the boiler was an old one, they were nowhere reduced in thickness by wear. The usual pressure of steam was 18 lbs., and a self-registering gauge showed that at the time of the explosion it did not exceed 20 lbs. The boiler was fitted with float, alarm whistle, glass gauge, safety-valve and self-acting feed apparatus, and the even water mark, about 1 foot above the top of the tube, showed the latter had worked regularly. There was a fusible plug which was placed over the fire-grate, and appeared to have commenced to give way.

The effect of the explosion was, that the top of the fire-grate, on the right side, had rent longitudinally, and the upper part of the shell, consisting of four rings of plates, and also the top of the fireplace, opened out and blew away to a considerable distance.

The front end became disengaged and also blew away. The bell-mouth of the tube was detached and blown to the front, and the tube which remained in the back part of the shell was collapsed upwards so completely that the bottom nearly touched the top, which latter retained its original shape.

There were a few signs of shortness of water, but nothing certain or trustworthy.

The cause of the explosion was most likely the intrinsic weakness of boilers of this shape, especially over the fire, where the top is only retained in its shape by numerous stays. The boiler had been very frequently repaired at this, the weakest place, and its strength had been thus so reduced as to make it unable to bear even a few pounds more than the ordinary working pressure.

The whistle was found to have been gagged by hemp, carefully inserted, so that there is ground for supposing there had been intentional unfair usage.

No. 3. *January 8th*, 1866. 1 *killed*, 1 *injured.*

The boiler was of the ordinary type used on small steamboats, and consisted of a cylindrical shell, about 5 feet diameter and 20 feet long, with flat front and hemispherical back, with an internal cylindrical fireplace, with a flue leading nearly to the back of the boiler, and returning nearly to the front, and then passing up through the steam space and through the top of the shell to the chimney. The usual working pressure was 14 lbs.

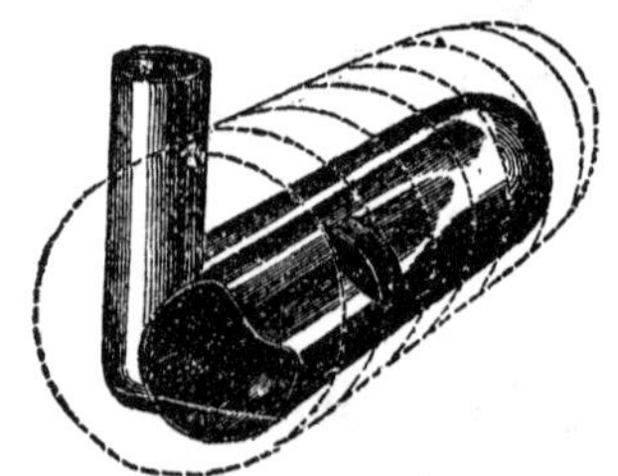

The effect of the explosion was, that a rent was made in the angle iron just within the fireplace, and the torn plate was bent downwards. There were altogether three rents in the tube over the fire, as a small band was held in its place by a stay from the tube to the shell. The escaping contents scalded those near, but no damage was done to anything but the boiler.

The cause was stated to be shortness of water. There was no gauge for ascertaining the height of the water.

No. 4. *January 17th*, 1866. 1 *injured.*

This was a boiler of the Cornish type, but no details have been obtained.

No. 5. *January 9th*, 1866. 1 *killed.*

This boiler was of the Cornish type, 28 feet long, 6 feet 6 inches diameter, with one tube 3–6 diameter, made of ⅜ inch plates, and worked at 45 lbs. The fittings were efficient. There was no sign of overheating.

The result of the explosion was that the tube collapsed in an inclined direction.

The cause was the weakness of the tube of such large diameter, which, without strengthening rings, was unable to resist the ordinary working pressure.

The appearance after explosion was nearly the same as that shown in sketch given at No. 12 explosion.

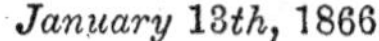

No. 6. *January 13th*, 1866. 4 *killed*, 4 *injured.*

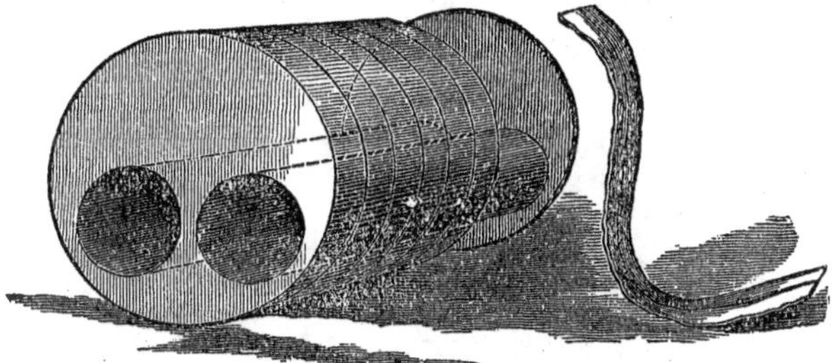

This was a two-tube Cornish boiler, 22 feet 3 inches long, and 7 feet 7 inches diameter, made of $\frac{3}{8}$ inch plates. The tubes were 2 feet 7 inches diameter, made of $\frac{3}{8}$ inch plates. The boiler was fitted with glass gauge and three gauge cocks and two safety-valves loaded to 40 lbs.

There was no sign of shortness of water.

The result of the explosion was, that about 8 feet of the back portion of the shell had torn off and blown away, leaving the tubes and ends intact.

The cause of the explosion was considered to be excessive pressure, but how accumulated did not transpire.

It is most natural to expect that there must have been corrosion under the bottom to cause the first rent.

No. 7. *January 18th*, 1866. 1 *killed*, 2 *injured.*

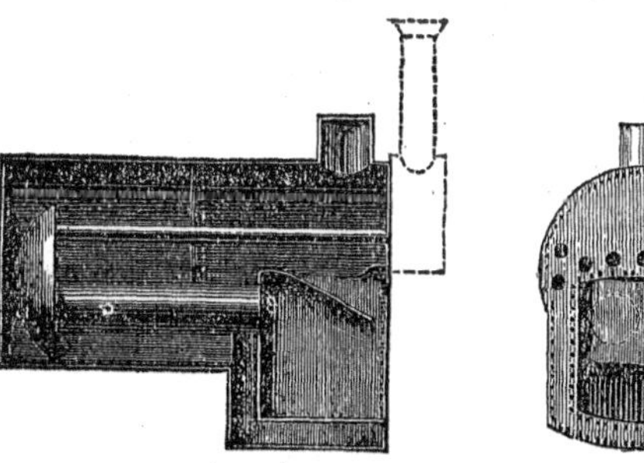

This boiler was of the agricultural type, 8 feet 2 inches long, 3 feet 6 inches diameter in the barrel, and made of $\frac{5}{16}$ inch plates. There was an internal fire-box 2 feet 4 inches wide, 2 feet 4 inches high, and 2 feet 10 inches deep. From the back of this fire-box two 12 inch tubes passed to an internal chamber at the back of the barrel. From this again passed nine $3\frac{3}{4}$ inch tubes to an exterior smoke-box fixed over the fire-door. The boiler was fitted with one spring safety-valve, which was screwed down tight, but had no pressure-gauge. The front plate was intended to be bolted on so that the fire-box and tubes could be taken out for cleaning, but it had been rivetted, leaving no means of cleaning, and it was nearly full of scurf.

The effect of the explosion was, that the crown-plate of fire-box, which was too flat and unstrengthened, rent at the front and along both top edges, and was driven into the fire. The boiler was thrown backwards and reared against a wall, resting on the front right-hand corner.

The cause of the explosion was the very dirty state, allowing of overheating where the water was not in proper contact with the plates, as the construction was so defective as not to admit of proper cleaning. The mountings were insufficient for the proper protection of the boiler.

No. 8. *January 29th*, 1866. 1 *killed*, 3 *injured.*

This was a plain cylinder boiler with hemispherical ends 30 feet long and 6 feet diameter, made of ⅜ inch plates, and worked at a pressure of 30 lbs. It was the second from the engine in a bed of fuse, and had been at work three years, then remained idle for eight years, working again fourteen years. The boiler was mounted with two safety-valves 3⅛ and 4 inches diameter, a float and alarm whistle. The boiler had been more than once repaired over the fire with several new plates. One seam over the fire had been observed to leak about a week before the explosion, but not seriously, and had been caulked. The boiler had just been started after cleaning. Six months before explosion, the boiler had been tested to 69 lbs.

The effect of the explosion was, that about 5 feet of the front end of the boiler opened out flat, and was thrown to the rear about 60 yards; the front hemispherical end was liberated and thrown 20 yards also to the rear and right hand. The back part of the boiler was thrown in a mass, and, after bounding twice, lodged at a distance of 230 yards.

The cause of the explosion appeared to be from the failure of a seam over the fireplace in a plate deteriorated by age, and overheated through a deposit of scurf and mud.

No. 9. *February 7th*, 1866. 1 *killed*, 4 *injured.*

This was a plain cylindrical boiler, with hemispherical ends, 23 feet long and 5 feet diameter, made of ⅜ inch (full) plates, and worked at 50 lbs. pressure. There was a small dome at the back. It was set so as to be fired under the bottom, if required, but the grate was seldom used The principal heat was supplied from a mill furnace, the neck of which was at the left hand of the back, and the flame was carried by a wheel flue round the front of the boiler to the stack on the right-hand side of the back. The boiler was fitted with a 4¾ inch safety-valve and also a float, but it was suspected that the latter had broken from the rod.

The effects of the explosion was, that a horizontal seam about the middle of the left-hand side had given way, and the upper portion of the third and fourth rings of plates had spread out like a lid, without being detached from the boiler. The front end became detached and was thrown some distance to the front.

The cause of the explosion was, that the left side of the boiler became overheated and so much softened, that it first bulged outwards with the ordinary

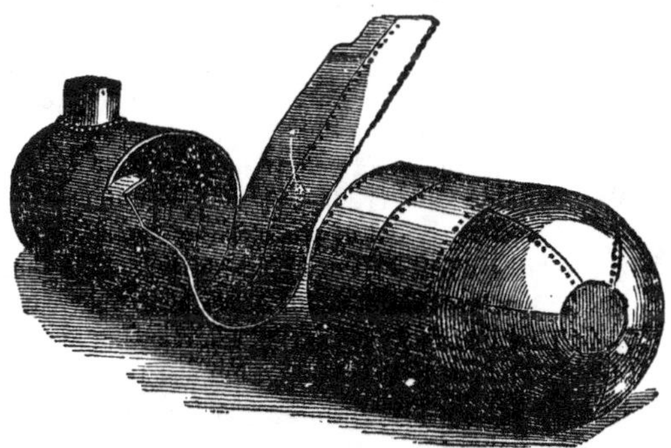

working pressure, and then rent open. The overheating was most likely the result of shortness of water, but it might possibly have been in consequence of the intense heat of a mill furnace impinging on a small area so near the water-line, leading to such rapid ebullition as to prevent sufficient contact of the water to keep the plates cool.

No. 10. *February 14th*, 1866. 1 *injured.*

This was a railway locomotive, and it exploded while standing in a shed, but no particulars have been obtained.

No. 11. *February 26th*, 1866. 1 *injured.*

This was a marine boiler in a small tug boat, but no details have been obtained.

No. 12. *February 26th*, 1866. 1 *injured.*

This boiler was of the Cornish type, 21 feet long, 5 feet diameter, with 1 tube 2 feet 11 inches in diameter, made of plates ⅜ inch thick. There were no strengthening rings on the tube. The boiler was second-hand, and had only worked a part of a day at its new position when it exploded.

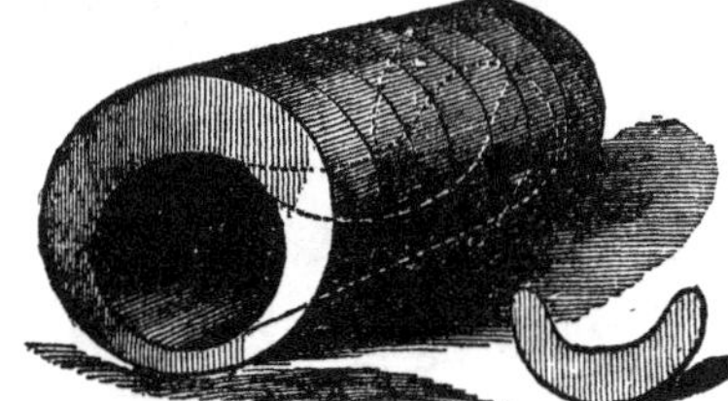

The pressure shortly before the explosion was 64 lbs. The tube collapsed, but the shell remained uninjured.

The cause of the explosion was the weakness of the tube without strengthening rings.

No. 13. *March 5th*, 1866. 7 *injured.*

This was a Cornish boiler worked at 27 lbs., but as the boiler was not injured, it is only noticed as an explosion, because of the imprudence of placing fittings in such dangerous positions. The boiler was beneath a work-room, and the lever of the safety-valve became displaced, allowing the valve to blow out, and the escaping steam rushed into the room above and scalded seven men very badly.

No. 14. *March 3d*, 1866. 1 *killed.*

This was a Cornish boiler, 31 feet 9 inches long, and 5 feet 9 inches diameter, with one tube 3 feet 8 inches diameter, made of $\frac{3}{8}$ inch plates, worked at 40 lbs. pressure.

The tube had been repaired with a bolted patch on the left side the day before the explosion, and it is presumed it was slightly out of the circular shape.

The effect of the explosion was, that the tube collapsed sideways, the top being thrown up above its proper height. There was no evidence of shortness of water.

The cause of the explosion was that the tube was too weak to sustain the ordinary working pressure, there being none of the strengthening rings found to be so essential to insure safety.

No. 15. *March 6th*, 1866. 1 *killed.*

This was a boiler with two internal fireplaces, connected into one flue at the back.

The effect of the explosion was, that the crowns of both furnaces were collapsed and slightly rent, and the escaping steam and water scalded the attendant.

The cause was that the water was 8 or 9 inches below its proper level, allowing the furnace crowns to become overheated and unable to bear the working pressure.

Each furnace had been fitted with a fusible plug, but they proved inefficient.

No. 1· *March 13th*, 1866. 1 *killed*, 1 *injured.*

This was a very small boiler, 8 feet long and 3 feet $2\frac{1}{2}$ inches diameter, with two small tubes through. It was fired externally. There was a dome on the top.

The effect of the explosion was, that the dome was blown off, and the rents continuing, the top opened out on each side, and the upper portion of both back and front was bent back.

The cause of the explosion was, that the dome cut away the whole of the top plate, and so much reduced the strength that it would not bear the ordinary working pressure.

No. 17. *March 19th*, 1866. *None injured.*

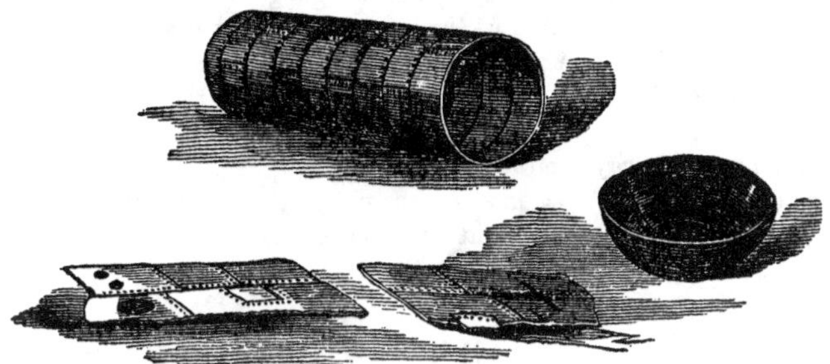

This was a plain cylinder boiler, 36 feet long, 5 feet 6 inches diameter, and worked at a pressure of 60 lbs. It had frequently been repaired over the fire, so that the longitudinal seams ran for several plates without break of joint. A patch had been put on a few days before explosion, and as the rivet holes had badly fitted, there had been much strain caused by drifting, and the rivets were much distorted.

The effect of the explosion was, that a longitudinal seam gave way over the fire, when two rings of plates opened out, rending the transverse seams at each side until completely separated, and fell in two parts at a distance of about 100 yards in front of the boiler. The front end was liberated, and fell in one piece about 100 yards beyond the two pieces of the shell. The main body of the boiler was driven back a few yards, and rolled over so as to be upside down, but was little injured.

The cause of the explosion was the frequent and badly executed repairs over the fireplace, which so weakened the structure as to make it unable to bear the very high ordinary pressure. This frequent repair over the fireplace had been made necessary by very hard firing and deposits of scurf from muddy water, preventing the proper contact of the water with the plates.

No. 18. *March 22d*, 1866. 1 *killed.*

This was a small boiler to supply steam for some steam winches on board a steamboat, and exploded through the overheating of the upper tubes through shortness of water, or from over-pressure in consequence of having only one safety-valve.

No. 19. *March 27th*, 1866. 2 *killed*, 18 *injured.*

This boiler was of the Cornish type, 24 feet 6 inches long, 6 feet 6 inches diamete- The tubes were 2 feet 6 inches diameter, made of ⅜ inch plates. The working pressure was, 54 lbs. It was fitted with a steam gauge, blow-off cock, a 1½ inch dead weight safety-valve, loaded to 54 lbs., and a 4⅝ inch lever safety-valve, loaded to 62 lbs. The boiler had been at work about five years.

The effect of the explosion was, that nearly the whole of the two back rings of the shell were torn off and opened out, and the boiler was turned partly round and moved upon its seat sideways and forwards.

The cause of the explosion was most extensive corrosion at the under side of the back of the shell, where the first rent took place, caused by the leaking of the joints and seams.

No. 20. *April 4th,* 1866. 5 *killed,* 4 *injured.*

This was a Cornish boiler, 30 feet long and 7 feet diameter, with one tube slightly oval about 4 feet diameter, made of $\frac{7}{16}$ inch plates, and worked at 43 lbs. pressure. The boiler was fitted with a 3½ inch safety-valve, which is much too small for such a boiler. There was also a glass water-gauge, two gauge-cocks and a pressure-gauge.

The effect of the explosion was, that the tube collapsed from end to end. The front end was blown out with a short length of the tube attached, and was driven against a wall about 30 yards to the front. The main body of the shell and the back end, with the collapsed tube within it, were driven back against another wall about the same distance away. The explosion was so violent that very great damage was done to the surrounding property. There was no evidence of shortness of water.

The cause of the explosion was the weakness of the tube of so large a diameter, without strengthening rings, which made it unable to bear the ordinary working pressure. It is very probable, however, that at the time of the explosion, the pressure was considerably more than usual during the sudden stoppage of the engine, and the confusion caused by a man becoming entangled in the machinery.

No. 21. *April 10th,* 1866. 1 *killed.*

This was a plain cylinder boiler, with hemispherical ends, 34 feet long and 5 feet diameter, made of ⅜ inch plates placed lengthways, and worked at 33 lbs. pressure. It was mounted with the usual fittings, and was not corroded.

The explosion took place at a seam in the front part of the bottom of the boiler, just over the fire. This rupture allowed the sides to expand, until the boiler was completely destroyed and torn into seven pieces.

The cause of the explosion was supposed to be the defective state of the seams over the fire, which, being placed longitudinally, were in the weakest position.

No. 22. *April 26th*, 1866. 2 *killed*, 2 *injured*.

This was a small internally fired boiler, 5 feet high, 2 feet 4 inches diameter, and was intended to work at 70 lbs. The fittings were defective, the spiral spring of the safety-valve being most easily altered so as to cause over-pressure. The manhole was not strengthened by any ring, and the first rents commenced at that point.

The cause of the explosion was over-pressure and defective construction.

The sketch at No. 57 explosion in this year is of a similar boiler, which exploded from nearly similar causes.

No. 23. *April 21st*, 1866. *None injured.*

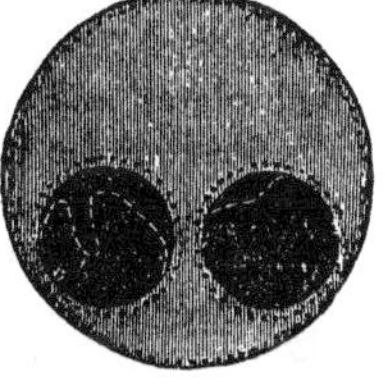

This was a Cornish boiler, 15 feet long, 6 feet diameter, with two tubes 1 foot 8 inches diameter, made of ⅜ inch plates, and worked at 40 lbs. pressure. The heat was supplied by two furnaces, one of which played into each tube.

The effect of the explosion was, that the left-hand tube collapsed, and tore away from the angle-iron in the front plate, and allowed the contents to issue violently and scatter the brickwork, but the boiler was not disturbed.

The cause of the explosion was shortness of water, as the collapsed tube had evidently been overheated.

No. 24. *May 13th*, 1866. *None injured.*

This was an old-fashioned balloon or haystack boiler, about 16 feet diameter, made of $\frac{5}{16}$ inch plates, and worked at very little above atmospheric pressure.

The boiler was chiefly used to store water during the time another boiler by the side of it was emptied. When the water was required to refill the other boiler, a fire was lighted under the balloon, and sufficient steam generated to drive the water from it into the other boiler. The safety-valve never being used, it had become fast, and as a little more steam than usual had accumulated, the bottom gave way, and the reaction of the issuing contents made the boiler rise from its seat, and it fell on its side at some distance away, flattened by the fall.

No. 25. *May 25th*, **1866.** ***None injured***

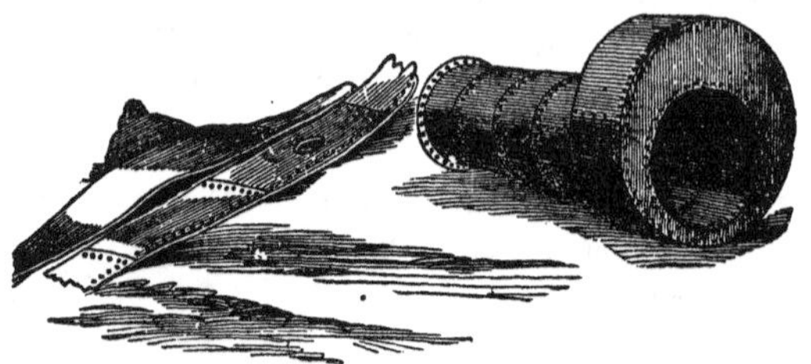

This was a Cornish boiler, 15 feet long and 4 feet 6 inches diameter, with one taper tube, 2 feet 9 inches diameter in front, and 2 feet diameter at back, made of $\frac{3}{8}$ inch plates, and worked at a pressure of 30 or 40 lbs. The boiler rested on two walls forming the bottom flue. There was a safety-valve, a glass water gauge, a pressure gauge, and two fusible plugs upon the tube. There was nothing to show the shortness of water or over-pressure.

The effect of the explosion was, that two longitudinal rents took place on the upper side of the shell, allowing two strips, forming the central part of the shell, to open out by the continuation of the fracture, until they were blown to a considerable distance. The tube, with the front end, and one ring of the shell were thrown to the front, while the back end was thrown to the rear, the smallest end of the tube having torn away from the back.

The cause of the explosion was, that the shell was deeply corroded where it rested on the side walls of the bottom flue, and the strength of the boiler was thereby so much reduced that it was unable to bear the ordinary working pressure.

No. 26. ***May*** **26*th*, 1866.**  **1 *killed.***

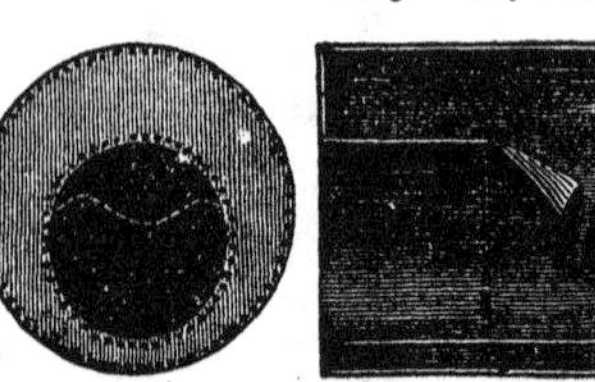

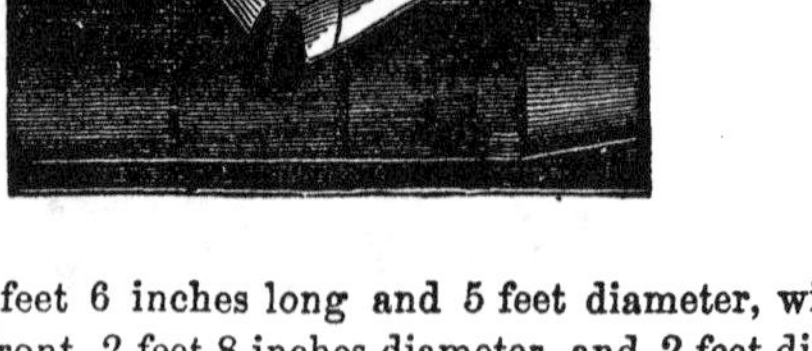

This was a Cornish boiler, 24 feet 6 inches long and 5 feet diameter, with one taper tube, slightly oval in the front, 2 feet 8 inches diameter, and 2 feet diameter at the back; made of $\frac{3}{8}$ inch plates. The boiler was fitted with a glass water gauge, and also a float, a self-acting feed apparatus, and a safety-valve loaded to 52 lbs., and also a mercury gauge.

The effect of the explosion was, that the tube collapsed over the fire, a rent taking place in the second ring of the plates. The issuing steam and water caused the death of a man in front, but the shell of the boiler was not injured or moved.

The cause of the explosion was shortness of water, and as the glass gauge was set unusually low, the man in charge may have been deceived. The oval shape of the fireplace, and the laminated iron, as shown in the fracture, rendered the tube peculiarly liable to collapse.

No. 27. *May 26th*, 1866. 1 *killed*

This was a plain cylinder boiler, 34 feet long, 5 feet diameter, made of ⅜ inch plates, and worked at 45 lbs. pressure. It was fitted with two 5 inch safety-valves and two floats.

The effect of the explosion was, that the boiler was torn in two pieces, that were thrown to a considerable distance. The first rent had taken place immediately over the fire.

The cause of the explosion was the weakening of the shell by the frequent repair over the fire, and this repair was rendered necessary by the deposit from the muddy water preventing proper contact of the water with the plates.

No. 28. *May 27th*, 1866. 1 *killed.*

This was a plain cylinder boiler, 32 feet long, 6 feet diameter, made of ⅜ inch plates, and worked at a pressure of 35 lbs. It was fitted with two safety-valves, two floats, and two alarm whistles.

The effect of the explosion was, that the boiler was lifted from its seat, and one end was separated and thrown to a considerable distance.

The cause of the explosion was weakening of the shell from repair a few days before, and perhaps over-pressure, as the gauge was found some time after, and indicated that the pressure had at some time exceeded 80 lbs.

No. 29. *May 28th*, 1866. 1 *killed*, 4 *injured.*

This was a Cornish boiler, 30 feet 8 inches long, 6 feet 8 inches diameter, with one tube, 4 feet diameter, made of $\frac{7}{16}$ inch plates, and worked at 40 lbs. pressure.

The effect of the explosion was, that the tube collapsed, and rent, and the issuing steam and water scalded those near.

The cause of the explosion was, that the tube was too weak to bear the ordinary working pressure, it having no strengthening rings.

No. 30. *May 31st*, 1866. 1 *killed*, 1 *injured.*

This was a plain cylindrical boiler, with dished ends, and only 4 feet 2 inches long and 2 feet 6 inches diameter, and made of ¼ inch plates. The boiler was most inefficiently mounted, the safety-valve was only 1⅝ inches diameter, and of such faulty construction, that it would not open under a pressure of 162 lbs. There was no steam-gauge or float; and the gauge-cocks were defective. There was no means of putting water in the boiler when there was a pressure of steam. The man-hole was very large for so small a boiler

The effect of the explosion was, that four rents starting from the manhole continued along the top of the boiler and round the end seams. A tongue-shaped strip of the top plate was attached to the back end plate; two strips about a foot wide on either side were blown away. The boiler had been turned nearly round in its flight, and fell with the back about 12 feet from the original position of the front.

The cause of the explosion was, that the boiler was worked until it was nearly dry, and during a temporary stoppage of the engine an accumulation of steam caused a greater pressure than the boiler could bear. The large manhole being the weakest place, the rent commenced there.

No. 31. *June 7th*, 1866. *None injured*

This was a marine boiler in a tug boat.

The effect of the explosion was, that the boiler was blown completely out of the vessel, and the greater part of it fell into the water, and a large piece alighted on a crowded quay, but without doing any damage.

The cause of explosion was supposed to be over-pressure during a temporary stoppage of the engine.

No. 32. *June 11th*, 1866. 2 *injured.*

This was a portable boiler of the agricultural type, of about 7 horse-power. The barrel of the boiler was 6 feet 1 inch long, 2 feet 5 inches diameter; the fire-box end was 3 feet wide and 2 feet 4 inches deep; the fire-box was 5 feet 5½ inches wide, and 2 feet 7 inches high, and 1 foot 9½ inches deep, with 23 tubes passing from it through the barrel to the smoke-box and chimney. The boiler was fitted with a 2 inch safety-valve, which was intended to blow at 45 lbs., but as there was no ferrule, it is supposed to have been screwed down to a much greater pressure.

The effect of the explosion was, that the upper portion of the shell over the fire-box rent through the manhole and allowed the shell to open out and fall on each side. A large portion of the front plate was also torn off.

The cause of the explosion was the weakness of the manhole, which was not strengthened by any ring, and also excessive pressure from want of proper safety-valve.

No. 33. *June 11th*, 1866. 1 *killed.*

This was a Cornish boiler, 36 feet 6 inches long, and 6 feet diameter, made of ⅜ inch plates, and worked at 45 lbs. pressure.

The effect of the explosion was, that the tube collapsed and rent, and the issuing contents caused the death of the attendant.

The cause of the explosion was the weakness of the tube of such large diameter, which was unable to bear the ordinary working pressure, having no strengthening rings.

No. 34. *June 19th*, 1866. 2 *killed*, 4 *injured.*

This was a railway locomotive, made of ½ inch plates, and worked at 140 lbs pressure.

The explosion occurred at the left-hand side of the ring of plates in the barrel next the fire-box and below the foot-plate. The rent tore along the edge of the lap and into the next ring of plates. The reaction of the issuing contents threw the engine off the rails.

The cause of the explosion was partial corrosion at the point of rupture and strain of the plates, as the boiler itself formed part of the frame of the engine.

No. 35. *June 26th*, 1866. 2 *injured*

This was a new railway locomotive, and was being tried for the first time, when the funnel came in contact with a bridge, and the dome was also torn off.

The cause of the explosion was, therefore, extraneous to anything belonging to the construction of the boiler.

No. 36. *June 26th*, 1866. *None injured*

No details have been obtained.

No. 37. *July 2d*, 1866. 4 *killed*

This was a plain cylindrical boiler, 30 feet long and 6 feet diameter, made of ⅜ inch plates, and worked at 28 lbs. pressure. The boiler had been repaired a short time before the explosion, with five new plates.

The effect of the explosion was, that the boiler was torn up into several pieces, but the main portion remained flattened out on the seating, while some smaller pieces were sent 350 yards away, but as the fragments were cleared away and cut up before proper examination could be made, it was impossible to trace out the first rent.

The cause of the explosion was most likely the deterioration of the boiler and its frequent repair over the fireplace during its nine years of working.

No. 38. *June 12th*, 1866. 4 *injured.*

This was an old-fashioned elephant boiler, 20 feet long and 4 feet diameter, made of ⅜ inch plates, and worked at low pressure. The bottom shell had a tube through its whole length.

The effect of the explosion was, that a rent took place in the forepart of the fire-place, and extending along the bottom, and the reaction of the issuing contents caused the top to rear up.

The cause of the explosion was supposed to be that the bottom plates were worn too thin to bear the ordinary pressure.

No. 39. *July 4th*, 1866. *None injured.*

This was a Cornish boiler externally fired, 30 feet long and 6 feet diameter, with two internal tubes made of ⅜ inch plates, and worked at 40 lbs. pressure.

The effect of the explosion, although not violent, was that the second seam over the fire gave way, and the plate sank down upon the fire.

The cause of the explosion was the deterioration of the seams over the fire, most likely in consequence of the deposit of scurf which could not be properly cleared off, owing to the internal tubes.

No. 40 *July 14th*, 1866. *None injured.*

This was a boiler with two internal furnaces, 9 feet 6 inches long and 2 feet 11 inches diameter, made of $\frac{3}{8}$ inch plates, uniting into one tube beyond.

The effect of the explosion was, that the furnace crown collapsed near the front of the boiler.

The cause of the explosion was, that there was an extra weight upon the safety-valve, and the steam-valve was left closed, so that more pressure accumulated than the boiler could bear.

No. 41. *July 23d*, 1866. 3 *injured.*

This was a rag boiler, and was not used for the generation of steam. It was a plain cylinder with hemispherical ends, about 16 feet long and 7 feet diameter. There was a neck at each end, upon which the boiler revolved, and through one of these the steam was admitted to a pressure of 30 lbs., in order to assist in cleaning the rags. There was a large manhole for filling and emptying.

The effect of the explosion was, that the boiler rent in the middle, and each half was blown to some distance.

The cause of the explosion was, that the manhole was so large that the strength of the boiler was too much reduced, and the constant strain of revolving caused a central seam to give way at the ordinary pressure.

The sketch to No. 63 is of a similar boiler.

No. 42. *July 28th*, 1866. 2 *killed*, 7 *injured.*

This was a boiler 36 feet 6 inches long, and 8 feet 9 inches diameter, made of $\frac{7}{16}$ inch plates, with flat back and hemispherical front end. A tube 3 feet 3 inches diameter passed from the back end nearly to the front, and returned to the back end, but of 6 inches less diameter, and then passed to an iron chimney. The fire-grate was beneath the hemispherical end. The working pressure was 36 lbs.

The effect of the explosion was, that the angle-iron of the flat back end gave way, and it separated from the shell and with the tubes attached, was blown to a good distance. The reaction drove the shell far in the opposite direction.

The cause of the explosion was the bad construction of the boiler, as the back end was quite unsupported, as there were no stays, and the bend of the tubes was not attached to the shell, and therefore the boiler was not strong enough to bear the ordinary working pressure.

An adjoining boiler, which was being cleaned inside by two men, was rolled off its seat by the force of the explosion.

No. 43. *August 7th*, 1866. 1 *killed*, 3 *injured*.

This was a railway locomotive, 13 feet 4 inches long, 3 feet 11 inches diameter, with 140 2 inch tubes. The fire-box was 4 feet 5 inches long, 3 feet 6 inches broad, and 5 feet deep, and made of copper ½ inch thick, and worked at 100 lbs. pressure. It was fitted with 2 4 inch safety-valves and a steam-gauge.

The effect of the explosion was, that the fire-box gave way about the middle of the left side, 2 feet 6 inches below the water-line, and the issuing water and steam scalded those near.

The cause of the explosion was, that the fire-box was corroded to ⅛ of an inch in thickness at the part which gave way.

No. 44. *August 2d*, 1866. 2 *killed*, 6 *injured*.

This was a plain cylindrical boiler, with flat ends. 23 feet long 5 feet 3 inches diameter, made of ⅝ inch plates, and worked at 40 lbs. pressure.

The effect of the explosion was, that both the flat ends were blown out, and the first ring of plates at the front end torn off.

The cause of the explosion was the weakness of the flat ends without stays.

No. 45. *August 31st*, 1866. 3 *killed*, 6 *injured*.

This was an upright boiler, 36 feet high 5 feet 6 inches diameter, made of $\frac{7}{16}$ inch plates, and worked at 45 lbs. pressure.

The effect of the explosion was, that the bottom gave way and the main bulk of the boiler was thrown straight up into the air to a great height, but descended again on its seating.

The cause of the explosion was attributed to shortness of water, but no particulars have been obtained.

No. 46. *August 22d*, 1866. 3 *killed*, 5 *injured*.

This was a marine multitubular boiler, 5 feet 8 inches long and 6 feet 6 inches diameter. There were two internal tubular furnaces, which joined to an internal chamber of large size, the small tubes passed to a smoke-box and chimney in the front over the fire-doors. Both ends were flat.

The effect of the explosion was, that the back was blown completely out and torn into two pieces, the lower portion remaining in the vessel, and the upper part falling in the water. The reaction of the issuing contents caused the boiler to be thrown on to the side of the quay.

The cause of the explosion was, that the flat back end was insufficiently stayed, and consequently unable to bear the ordinary pressure.

No. 47. *August 25th*, 1866. 1 *injured.*

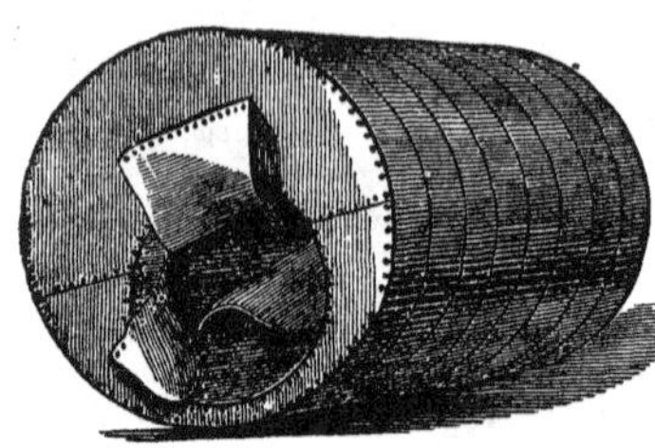

This was a Cornish boiler, 24 feet 3 inches long, and 6 feet diameter, with taper tube 3 feet 5 inches diameter, for about 7 feet 6 inches in length, and 2 feet 6 inches diameter for the rest of the length, and worked at 33 lbs. pressure.

The effect of the explosion was, that the fireplace gave way on the left side, and was so much torn that the plates were forced out of the front.

The cause of the explosion was, that the tube was so much corroded as to be too thin to bear the working pressure.

No. 48. *August 27th*, 1866. 3 *killed*, 1 *injured.*

This was a marine boiler of the usual construction, and had been tested to 60 lbs.

The effect of the explosion was, that it gave way at the lower portion of the back, and the issuing steam and water scalded those near.

The cause of the explosion was a seam rent, 6 feet 6 inches long, which had not been detected by testing.

No. 49. *August 29th*, 1866. 2 *killed*, 2 *injured.*

This was a portable agricultural boiler, and burst during a temporary stoppage from accumulation of steam, causing undue pressure; but no particulars have been obtained.

No. 50. *September 7th*, 1866. 2 *killed*, 30 *injured.*

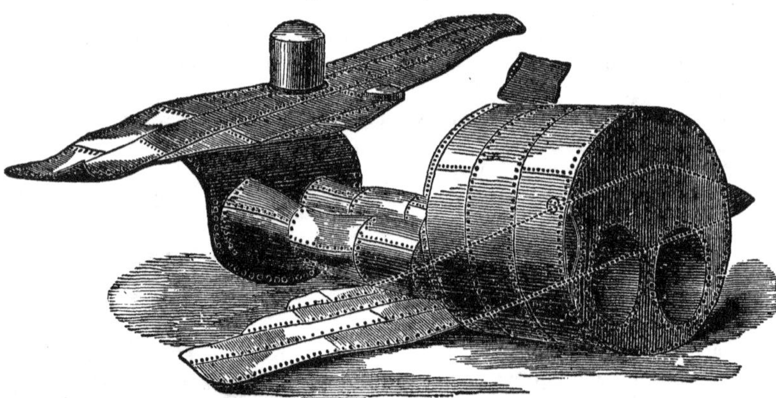

This was a Cornish boiler, 22 feet long, 7 feet 6 inches diameter, with 2 internal tubes, 3 feet diameter, made of $\frac{7}{16}$-inch plates, and worked at 60 lbs. pressure. There was a dome at the back part of the shell. The mountings of the boiler were efficient.

The effect of the explosion was, that some rents took place at the under side of the shell, allowing the central portion to open out and blow away. The portion containing the dome was thrown to the left, and the other to the right. The front end, with three rings of the shell, with the tubes and back end, were but little moved from their original position. The tubes were dented in on the top and bottom, by the fall of some large coping stones upon them, but the crowns of the furnaces were uninjured, and there was no sign of shortness of water or overheating.

The cause of the explosion was extensive corrosion on the under side of the shell, where it rested on the brickwork, and so reduced the strength that it was unable to bear the working pressure.

In the sketch the fragments are drawn so as to show their position when in the boiler.

No. 51. *September 25th*, 1866. 7 *killed.*

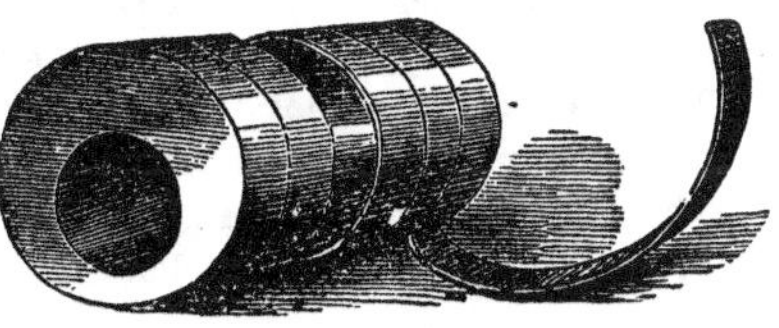

The boiler was of the Cornish type, and was 20 feet long, 4 feet 6 inches diameter, with a tube 2 feet 6 inches diameter. It was double riveted, and the crown of tube was strengthened with angle-iron. The plates were $\frac{3}{8}$ inch thick. The shell was formed of six rings, each of two plates alternately jointed top and sides. The third ring from the front had stripped off, and was thrown to the right and forwards against the wall. The line of rent was continued to the plates forming the ring, which was an outer one, and covered the two adjoining rings in the laps, the rent being from the edge of the inner lap to the nearest rivets. The first rent had taken place in the solid iron about 1 inch from the rivets of a seam on one side, and from this the rent had extended along the seams on either side, and of course the whole ring soon tore off when the equilibrium was destroyed by the first rent.

The fittings of the boiler were sufficient, except that there was only one safety-valve, and that was so constructed that it could only open a very little way.

The cause was a defect in the iron at the point of first rent, and accumulated pressure during the time of standing.

No. 52. *September 24th*, 1866. *None injured.*

This was a multitubular boiler with large internal fireplace, and worked at 60 lbs. pressure.

The effect of the explosion was, that the whole of the furnace crown was crushed down and torn across two seams. The boiler was lifted from its seat and thrown back against a stone wall.

The cause of the explosion was shortness of water, which allowed the furnace crown to become overheated.

No. 53. *September 23d*, 1866. 1 *injured.*

This was a plain cylindrical boiler, 7 feet long and 2 feet diameter, made of $\frac{3}{8}$ inch plates, and worked at 30 lbs. pressure.

The effect of the explosion was, that the upper part of the boiler at the first ring of plates was torn off, and the front end was blown out.

The cause of the explosion was extensive external corrosion, where the plates rested against the brickwork.

No. 54. *September* 21*st*, 1866. *None injured*

This was a Cornish boiler with one tube.

The effect of the explosion was, that the tube collapsed and parted, and the escape of the steam and water blew off the door-frame.

The cause of the explosion was supposed to be shortness of water, but no details have been obtained.

No. 55. *October* 5*th*, 1866. 1 *killed*, 7 *injured.*

This was a boiler of the ordinary agricultural type, and worked at about 45 lbs. pressure. The engine had only just been set to work.

The effect of the explosion was, that the crown-plate to the fire-box gave way, and the issuing contents scalded those near.

The cause of the explosion was, that the crown-plate was so deeply corroded from long wear, that it was incapable of bearing even the ordinary working pressure.

No. 56. *October* 8*th*, 1866. 2 *killed*, 2 *injured.*

This was a marine boiler, and only intended for low pressure. While the vessel was waiting to start, with steam up, the wing furnace of the starboard boiler collapsed on the wing side, as shown by the dotted lines, and allowed the steam and water to escape into the stokehole.

The boiler was slightly oval, and 16 feet long. The front end was flat, 8 feet 6 inches wide and 7 feet 10 inches high, and the dimensions of the back hemispherical end were 2 feet less each way. The plates were $\frac{3}{8}$ inch thick. The pressure was about 26 lbs. There were two internal fire-places of irregular shape, uniting at the back into one flue of similar shape, which did not come to the front, but passed through the steam-space and out at the top of the boiler.

The effect of the explosion was, that the side of the furnace next the shell was rent along the edge of a longitudinal seam in a line, which was slightly nicked in the caulking. This rent extended about 5 feet 6 inches from the front, and then at a cross seam it went along the line of rivets from the crown to the bottom of the furnace. Beyond this cross seam the furnace was collapsed, until it nearly touched the other side of the furnace, and the bulge died away towards the back end. There was also a rent in the lower part of the front of the shell, as shown in dotted line.

The cause of the explosion was the weakness of the shape of the flue, which was not stayed to the shell. It had evidently gradually been giving way some time before the explosion, and eventually collapsed at nearly the ordinary pressure. Symptoms of the same alteration of shape were noticed in the corresponding flue of the other boiler.

No. 57. *October 9th*, 1866. 7 *killed*, 1 *injured.*

The boiler was 5 feet 6 inches high and 2 feet 6 inches diameter, and had an internal conical fire-box, with two cross tubes and a chimney at the top, and was attached to a portable steam crane on board a sailing vessel.

The effect of the explosion was, that the outer shell of the boiler was rent into many pieces, leaving the central conical fireplace intact. The nature of the rents showed that the plate round the manhole, which was unstrengthened by a ring, had first given way, and all the other fractures had led away from that point. This is confirmed by the fact that the man-lid was thrown a good distance, with force enough to make its way through the timber walls of a cabin. The front plate divided into many pieces, and scattered right and left, while the back plate was thrown through a cabin in the opposite direction to the manhole.

The central flue showed a slight indication of over-heating, but the construction was such that the upper portion passed through the steam-space, and was always exposed to the action of the fire, without the protection of the contact of water. The shell of the boiler was only $\frac{1}{4}$ inch thick, and as the manhole was without a ring on its edge to strengthen the plate, and was held in by two clamps, which caused additional strain when carelessly screwed up, the manhole was by far the weakest place. The engine was standing after a short time of working, and as the safety-valve was very defective, and could be screwed down until tight, against almost any pressure, it is most probable that the pressure mounted much beyond the usual working pressure of 75 lbs., when the weakest part gave way and led to the sudden tearing up and scattering of the whole fabric.

The cause, therefore, was faulty construction of both boiler and fittings. The boiler being unable to bear that accumulated pressure, which the safety-valve ought to have made impossible.

No. 58. *October 13th*, 1866. 1 *killed.*

This was a Cornish boiler, 14 feet long, 6 feet diameter, with an internal flue 3 feet 3 inches by 2 feet 10 inches, made of $\frac{3}{8}$ inch plates, and worked at a pressure of 27 lbs. It stood on three saddles, with a bearing surface of 3 feet by 4 inches.

The effect of the explosion was, that a portion of plate 20 inches by 18 inches, at the bottom of the boiler, was blown out, and the contents scalded the man who was killed.

The cause of the explosion was, that the plate around this rupture was corroded to only $\frac{1}{16}$ inch.

No. 59. *November 1st*, 1866. 7 *killed.*

In this case two boilers exploded simultaneously, and both were on board a small steamboat. They were 16 feet long and 6 feet 6 inches diameter at the flat front ends, and somewhat less at the hemispherical back ends. Each had two internal fireplaces, which united in one flue, which returned nearly to the front and passed up through the steam space, and out at the top of shell into the funnel. The central fireplaces were not circular, and the outside fireplaces and the return flues were still more distorted, but the weakness of the shape was somewhat compensated for by stays between the tubes, and from the tubes to the shell. The mountings to the boiler were of the ordinary kind, and efficient.

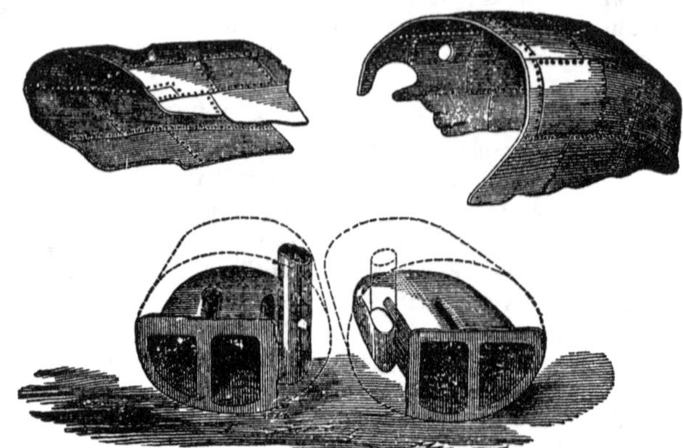

The effect of the explosion was, that the under side of the shell of each boiler was rent longitudinally for its whole length, allowing the sides to open out and tear away from the front end. They fell at a considerable distance. The furnaces, with parts of the front ends attached, fell into the water, but with the exception of one of the side flues, which was a little collapsed, they were uninjured. Small pieces of the front ends were thrown to a great distance. The sides of the vessel were so completely blown out that she sank. There were no signs of shortness of water or overheating.

The cause of the explosion was such deep corrosion on the under sides of the boilers, that very little of the iron was left, so that they were unable to bear the slightly increased pressure from a temporary stoppage of the engine.

This corrosion was, no doubt, caused by the leakage of the vessel keeping the shells of the boilers constantly wet.

No. 60. *November 3d*, 1866. *None injured.*

This boiler was of the locomotive or agricultural type; the barrel was 3 feet 9 inches long and 2 feet 6 inches diameter. There was an internal fire-box, and the heat passed through a number of 1 inch tubes to the front smoke-box and chimney. The usual working pressure was 70 lbs.

The explosion happened during a stoppage for breakfast time, and the effect was that the fire-box end was torn from the barrel, and from the position of those fragments that could be found, the boiler appeared to have turned over, as part of the fire-box was sent through the stage upon which the boiler was traveling, and the barrel with the tubes remaining in it first struck a rail, which caused it to be dented in, and then it rebounded to a point about 140 yards from its original position.

There was not enough of the fragments recovered from the river, into which they fell, to trace the cause of the explosion, but it is presumed that, although when the boiler was left, only 20 lbs. pressure was shown by the gauge, the fire-door being left closed, the pressure must have risen to a point much above the working pressure, and to more than the boiler could bear.

No. 61 *October* 1*st*, 1866. *None injured.*

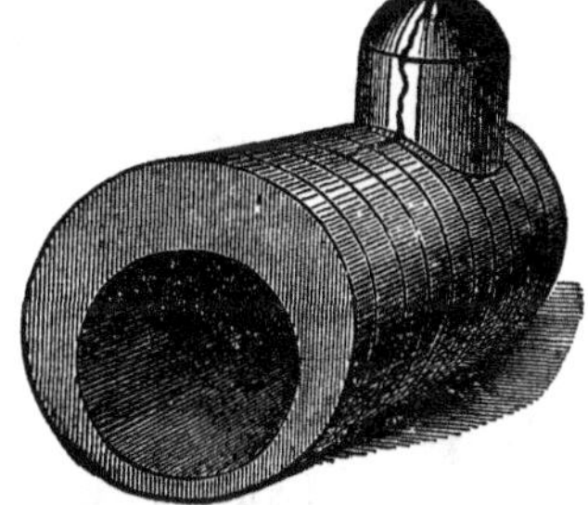

The boiler was of Cornish type, and was 22 feet long, 6 feet diameter, with one tube 4 feet 6 inches diameter, and made of $\frac{3}{8}$ inch plates. The working pressure was about 12 lbs. There was an unusually large dome at the back end, 5 feet diameter, and the whole of the shell was cut out from under it, so that the construction was peculiarly weak.

The boiler had been off for cleaning, and steam was being got up in the night, and it was said that an extra pressure was caused by the stop-valve being left closed, but it could not have been very great, or the large tube would have collapsed.

The effect of the explosion was, that the dome was split in two, in the line marked in the sketch, and the shell was a little ruptured on each side of it, and so large a rent was suddenly made, that the contents of the boiler passed harmlessly into the air without moving the boiler on its seat.

The cause was the extreme weakness of the shell at the juncture of the dome.

No. 62. *November* 11*th*, 1866. 1 *killed*, 1 *injured.*

FRONT VIEW. BACK VIEW.

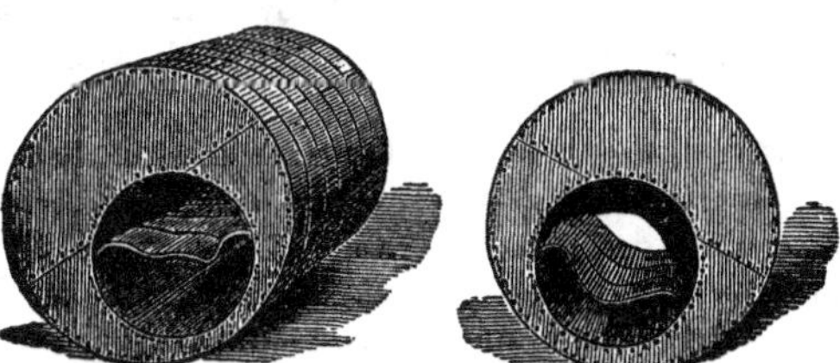

The boiler was of Cornish type, 16 feet 3 inches long and 5 feet 8 inches diameter, with one tube 3 feet 1 inch diameter, and was made with $\frac{7}{16}$ inch plates. The ordinary working pressure was 60 lbs.

The effect of the explosion was, that the tube collapsed from end to end. The top had fallen and the bottom had risen until they nearly met. The tube was torn away from the angle-iron for the upper half of the circle at both ends. The shell of the boiler was uninjured, and the front and back flat ends were only slightly bulged outwards. The boiler had only moved about a yard backwards, and a little on one side. The discharge of the contents of the boiler from either end had knocked down the walls of the premises, and moved some machinery, and let down much roofing.

There was no indication of shortness of water, or overheating of the plates; but the tube had collapsed from over-pressure, the man in charge having fastened three bricks to the lever of the safety-valve, under the impression that he could thus accumulate a quantity of steam during the night, to be available on commencing work on the morrow, but he was killed by the explosion.

The tube had no strengthening hoops, and was more than an inch out of the true circle at the ends, and was doubtless much more out of proper form in the centre, so that the extra pressure of the 27 lbs. weight at the end of the safety-valve lever caused it to collapse.

No. 63. *November 29th*, 1866. *None injured.*

This was not properly a boiler, as it was not exposed to the fire, but a revolving steam chamber. It was a plain cylinder with hemispherical ends, 12 feet 6 inches long and 5 feet diameter.

The manhole was large, to facilitate filling and emptying, and was rectangular and unstrengthened on the edges, and measured 8 feet 6 inches in length and 1 foot 6 inches in width. The lid fitted on the inside, and was held by clamps.

The boiler was much out of repair, and a crack shown in sketch, from one corner of the manhole to the commencement of the hemispherical end, was only temporarily prevented from leaking by a screw-patch, which did not restore the strength.

The explosion happened when, in revolving, the manlid was downwards, and the result was, that the lid was driven nearly through the floor, and the shell was rent from the opposite corners of the manhole and opened out, and the reaction of the escaping contents sent the boiler through the roof, and in descending, it fell through another roof into a floor at the ground level. The rent on the side originally cracked, had extended through the hemispherical end, dividing it into two halves; but on the other side the rent had followed the seam joining the shell to the end, and had run completely round the boiler, allowing the end to be blown away.

The cause of the explosion was the extreme weakness of the construction, as the large manhole cut away nearly all the strength on one side, and the fastenings o the lid were not contrived so as to compensate at all for the loss of strength. The constant strain upon the structures, when revolving, also tended to weaken it. These two causes together were sufficient to account for the explosion, at the usual working pressure of 12 lbs., although it is possible that it might have been increased to 35 lbs., as that was the pressure in the boiler supplying the steam, although the pressure was regulated by a check-valve.

This explosion (and also No. 41 in this year) will clearly show that a mere vessel of steam not exposed to the fire, or any chance of over-heating of the plates, can burst and cause very great destruction, although there could be no sudden *increase of pressure*, which is so often supposed necessary to account for the havoc caused by explosions.

No. 64 *December* 1*st*, 1866. 3 *killed*, 2 *injured.*

The boiler was of the agricultural portable class, and was 7 feet 6 inches long and 3 feet 8 inches diameter, made of ¼ inch plates. The fire-box was 2 feet long and 2 feet 10 inches broad. Two 11½ inch tubes led from it to an internal chamber at the other end of the boiler, and three 8¼ inch tubes led back again to an external smoke-box and chimney fixed over the fire-door. The boiler was mounted with a water-gauge, a steam-gauge and a spring safety-valve loaded to 35 lbs. The wheels had been removed.

The effect of the explosion was, that the bottom of the right-hand side lower tube was forced upwards, and rent along to within 12 inches of the fire-box.

The cause of the explosion was that the tubes which had been $\frac{3}{16}$ inch thick, were corroded so much from leakage from patches in the smoke-box, that only $\frac{1}{16}$ of an inch of the metal was left, so that they were unable to bear the slight increase of pressure that had accumulated during a short stoppage.

No. 65. *December* 4*th*, 1866. 2 *killed*, 6 *injured.*

This boiler was of the Cornish type, 22 feet long and 7 feet 6 inches diameter, with two tubes 3 feet diameter each, and they were strengthened with rings in the approved manner. The boiler altogether was strongly made, and less than three years old.

The effect of the explosion was, that the second from the back of the seven rings was ruptured at the bottom and torn off by a rent through the line of rivets on each side, and the boiler was thrown from its seat and turned completely over so as to lie in a contrary direction to what it was before. The ring torn off was opened out flat.

The cause was extensive corrosion, which resulted from leakage of the seams beneath the brickwork, and hidden from view.

No. 66. *December* 7*th*, 1866. *None injured.*

This boiler was a plain cylinder with hemispherical ends, and was about 9 feet long and 3 feet 8 inches diameter, and was worked at something under 20 lbs. pressure. It was set in the usual way, with a fire under one end, and a flash flue

The effect of the explosion was, that the boiler was rent all along one side, and the shell opened out, and the reaction of the issuing contents caused the shell to be thrown some yards away. One of the ends became altogether detached, and flew to a considerable distance.

The cause of the explosion appeared to be, that the plates had so much thinned by corrosion, that they were scarcely $\frac{1}{16}$ inch thick, and therefore unable to bear safely the ordinary working pressure, and gave way on a very slight increase of pressure during a temporary stoppage of the engine.

No. 67. *December* 16*th*, 1866. 1 *killed*, 1 *injured.*

No further particulars have been obtained, than that this boiler exploded through its being wasted by long wear and over-pressure.

No. 68. *December 15th*, 1866. 5 *injured*.

This was a Cornish boiler, 24 feet long, 6 feet 6 inches diameter, with two flues, 2 feet 8 inches diameter, and made of $\frac{7}{16}$ inch plates, and worked at 50 lbs. pressure. The boiler was fired in each of the tubes in the ordinary way, and also the heat from two furnaces passed from the back, one on each side of the outside shell.

The effect of the explosion was, that both the internal furnaces were collapsed, until the crowns almost touched the fire-bars, as shown in dotted lines, but without

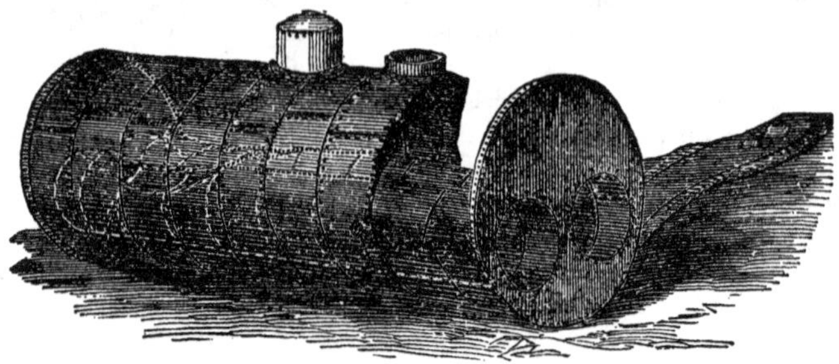

fracture. The back of the shell, on the right side, had evidently been overheated, and had rent along the centre of a bulge, and this rent had extended along the line of rivets of the transverse seam on each side, allowing two rings of the plates of the shell to open out flat, as shown. There was a bulge on the plate, on the right side of shell, corresponding with the one which parted on the opposite side.

The cause of the explosion was overheating of the plates from shortness of water. As the glass gauge showed plenty of water just before explosion, it was presumed that the bottom stop-cock was stopped up.

No. 69. *December 24th*, 1866. 1 *injured*.

No particulars have been obtained.

No. 70. *November 26th*, 1866. *None injured*.

This boiler was 28 feet long, 7 feet diameter, made of $\frac{7}{16}$ inch plates, and worked at 50 lbs. pressure. There were two internal fireplaces united into one tube beyond.

The effect of the explosion was, that the sides of the oval chamber forming the junction of furnaces and tube crushed inwards.

The cause of the explosion was, that this chamber was of such weak shape as to be unable to resist the ordinary working pressure.

www.ingramcontent.com/pod-product-compliance
Lightning Source LLC
LaVergne TN
LVHW021430110826
845150LV00007B/2172

* 9 7 8 1 4 2 5 5 0 6 5 6 8 *